Alfred Payson Gage

Introduction to Physical Science

Alfred Payson Gage

Introduction to Physical Science

ISBN/EAN: 9783337036102

Printed in Europe, USA, Canada, Australia, Japan

Cover: Foto ©berggeist007 / pixelio.de

More available books at **www.hansebooks.com**

INTRODUCTION

TO

PHYSICAL SCIENCE.

BY

A. P. GAGE, Ph.D.,

Instructor in Physics, English High School, Boston, and
Author of "Elements of Physics."

BOSTON:
GINN & COMPANY.
1889.

EDUCATION LIBR.

Entered according to Act of Congress, in the year 1887, by
A. P. GAGE,
in the Office of the Librarian of Congress, at Washington.

TYPOGRAPHY BY J. S. CUSHING & CO., BOSTON.
PRESSWORK BY GINN & CO., BOSTON.

AUTHOR'S PREFACE.

An experience of about six years in requiring individual laboratory work from my pupils has constantly tended to strengthen my conviction that in this way alone can a pupil become a master of the subjects taught. During this time I have had the satisfaction of learning of the successful adoption of laboratory practice in all parts of the United States and the Canadas; likewise its adoption by some of the leading universities as a requirement for admission. Meantime my views with reference to the trend which should be given to laboratory work have undergone some modifications. The tendency has been to some extent from qualitative to quantitative work. With a text-book prepared on the inductive plan, and with class-room instruction harmonizing with it, the pupil will scarcely fail to catch the spirit and methods of the investigator, while much of his limited time may profitably be expended in applying the principles thus acquired in making physical measurements.

A brief statement of my method of conducting laboratory exercises may be of service to some, until their own experience has taught them better ways. As a rule, the principles and laws are discussed in the class-room in preparation for subsequent work in the laboratory. The pupil then enters the laboratory without a text-book, receives his note-book from the teacher, goes at once to any unoccupied (numbered) desk containing apparatus, reads on a mural blackboard the questions to be answered, the directions for the work to be done with the apparatus, measurements to be made, etc. Having performed the necessary manipulations and made his observations, he surrenders the apparatus to another who may be ready to use it, and next occupies himself in writing up the results of his experiments in his note-book. These note-books are deposited in a receptacle near the door as he leaves the laboratory. Nothing is ever written in them except at the times of experimenting. These books are examined by the teacher; they contain the only written tests to which the pupil is subjected, except the annual test given under the direction of the Board of Supervisors. Pupils, in general, are permitted to communicate with their teacher only. "Order, Heaven's first law," is absolutely indispensable to a proper concentration of thought and to successful work in the laboratory.

Only in exceptional cases, such as work on specific gravity and electrical measurements, has it been found necessary to duplicate apparatus. The same apparatus may be kept on the desks through several exercises, or until every pupil has had an opportunity of using it. Ordinarily two pupils do not perform the same kind of experiment at the same time. With proper system, any teacher will find his labors lighter than under the old elaborate lecture system; and he *will never have occasion to complain of a lack of interest on the part of his pupils.*

I venture to hope, in view of the kind and generous reception given to the Elements of Physics, that this attempt to make the same methods available in a somewhat more elementary work may prove welcome and helpful. It has been my aim in the preparation of this book to adapt it to the requirements and facilities of the average high school. With this view, I have endeavored to bring the subjects taught within the easy comprehension of the ordinary pupil of this grade, without attempting to "popularize" them by the use of loose and unscientific language or fanciful and misleading illustrations and analogies, which might leave much to be untaught in after time. Especially has it been my purpose to carefully guard against the introduction of any teachings not in harmony with the most modern conceptions of Physical Science.

I would here acknowledge, in a very particular manner, my obligations for invaluable assistance rendered by Dr. C. S. Hastings, Professor of Physics in the Sheffield Scientific School, New Haven, Conn., and Prof. S. W. Holman, of the Massachusetts Institute of Technology, both of whom have carefully read all the proof-sheets. It would, however, be highly improper to attribute to them in any measure responsibility for whatever slips or inaccuracies may have crept into these pages. I am also under obligations for valuable suggestions and criticisms received from the veteran educator, Prof. B. F. Tweed, of Cambridge, Mass.; George Weitbrecht, High School, St. Paul, Minn.; John F. Woodhull, Normal School, New Paltz, N.Y.; Robert Spice, Professor of Physics in the Technological Institute, Brooklyn, N.Y.; C. Fessenden, High School, Napanee, Ont.; A. H. McKay, High School, Pictou, N.S.; and F. W. Gilley, High School, Chelsea, Mass.

CONTENTS.

CHAPTER I.

Matter, energy, motion, and force. — Attraction of gravitation. — Molecular and molar forces 1

CHAPTER II.

Dynamics of fluids. — Pressure in fluids. — Barometers. — Compressibility and elasticity of gases. — Buoyancy of fluids. — Density and specific gravity 29

CHAPTER III.

General dynamics. — Momentum, and its relation to force. — Three laws of motion. — Composition and resolution of forces. — Center of gravity. — Falling bodies. — Curvilinear motion. — The pendulum 67

CHAPTER IV.

Work and energy. — Absolute system of measurements. — Machines 98

CHAPTER V.

Molecular energy, heat. — Sources of heat. — Temperature. — Effects of heat. — Thermometry. — Convertibility of heat. — Thermo-dynamics. — Steam-engine 121

CONTENTS.

CHAPTER VI.

                       PAGE

Electricity and magnetism. — Potential and electro-motive force. — Batteries. — Effects produced by electric current. — Electrical measurements. — Resistance of conductors. — C.G.S. magnetic and electro-magnetic units. — Galvanometers. — Measuring resistances. — Divided circuits; methods of combining voltaic cells. — Magnets and magnetism. — Current and magnetic electric induction. — Dynamo-electric machines. — Electric light. — Electroplating and electrotyping. — Telegraphy. — Telephony. — Thermo-electric currents. — Static electricity. — Electrical machines 154

CHAPTER VII.

Sound. — Study of vibrations and waves. — Sound-waves, velocity of; reflection of; intensity of; reënforcement of; interference of. — Pitch. — Vibration of strings. — Overtones and harmonics. — Quality. — Composition of sonorous vibrations. — Musical instruments. — Phonograph. — Ear 238

CHAPTER VIII.

Radiant energy, ether-waves, light. — Photometry. — Reflection of light-waves. — Refraction. — Prisms and lenses. — Prismatic analysis. — Color. — Thermal effects of radiation. — Microscope and telescope. — Eye. — Stereopticon 281

APPENDIX: A, metric system; B, table of specific gravity; C, table of natural tangents; D, table of specific resistances . 341

INDEX 349

"*Nature is the Art of God.*" — THOMAS BROWNE.

CHAPTER I.

MATTER, ENERGY, MOTION, AND FORCE.

Section I.

MATTER AND ENERGY.

TO THE TEACHER: — *That portion of this book which is printed in the larger type, including the experiments, is intended to constitute in itself a tolerably full and complete working course in Physics.* The portion in fine print may, therefore, be wholly omitted without serious detriment; or parts of it may be studied at discretion as time may permit; or, perhaps still better, it may be used by the student, in connection with works of other authors, as *subsidiary reading.* It should be borne in mind that recitations from memory of mere descriptive Physics and Chemistry is of little educational value.

TO THE PUPIL: — "Read nature in the language of experiment"; that is, put your questions, when possible, to nature rather than to persons. Teachers and books may guide you as to the best methods of procedure, but your own hands, eyes, and intellect must acquire the knowledge directly from nature, if you would *really know.*

1. Matter. — *Physics* including Chemistry, may for the present purposes at least be regarded as *the science of matter and energy.* The question, What is matter? is *apparently* a very simple one, and easy to answer. One of the first answers that will occur to many is, Anything that can be seen is matter.

2. Is Matter ever Invisible? — We are usually able to recognize matter by seeing it. We wish to ascertain by experiment, *i.e.* by putting the question to nature, whether matter is ever invisible. Now in experimenting there must (1) be certain facts of which we are tolerably certain at the outset. These facts (2) lead us to place things in certain situations (the operation is called *manipulation*) in order to ascertain what results will follow. Then, in the light of these results we (3) *reason from the things previously known to things unknown, i.e.* to facts which we wish to ascertain.

For example, we are certain that we cannot make our two hands occupy the same space at the same time. All

Fig. 1.

Fig. 2.

experience has taught us that *no two portions of matter can occupy the same space at the same time.* This property (called *impenetrability*) of *occupying* space, and not only occupying space, but *excluding* all other portions of matter from the space which any particular portion may chance to occupy, is peculiar to matter; nothing but matter possesses it. This known, we have a key to the solution of the question in hand.

There is something which we call air. It is invisible. *Is air matter?* Is a vessel full of it an "empty" vessel as regards matter?

Fig. 3.

Experiment 1.— Thrust one end of a glass tube to the bottom of a basin of water; blow air from the lungs through the tube, and watch the ascending bubbles. Do you see the *air* of the bubbles, or do you see certain *spaces* from which the air has excluded the water?

Is air matter? Is matter ever invisible? State clearly the argument by which you arrive at the last two conclusions.

Experiment 2. — Float a cork on a surface of water, cover it with a tumbler (Fig. 1) or a tall glass jar (Fig. 2), and thrust the glass vessel, mouth downward, into the water. (In case a tall jar is used, the experiment may be made more attractive by placing on the cork a lighted candle.) State what evidence the experiment furnishes that air is matter.

Relying upon the impenetrability of air, men descend in diving-bells (Fig. 3) to considerable depths in the sea to explore its bottom, or to recover lost property.

Fig. 4.

Observe the cloud (Fig. 4) formed in front of the nozzle of a boiling tea-kettle. All the matter which forms the large cloud escapes from the orifice, yet it is invisible at that point, and only becomes visible after mingling with the cold outside air. Place the flame of an alcohol lamp in the cloud; the matter again becomes nearly or quite invisible in vicinity of the flame. *True steam is never visible.* Here we see matter undergoing several changes from the visible to the invisible state, and *vice versa*.

3. Matter, and only Matter, has Weight. — *Has air weight?*

Experiment 3. — Suspend from a scale-beam a hollow globe, *a* (Fig. 5), and place on the other end of the beam a weight, *b* (called a counterpoise), which just balances the globe when filled with air in its usual condition. Then exhaust the air by means of an air-pump, or (if the scale-beam is very sensitive) by suction with the mouth. Having turned the stop-cock to prevent the entrance of air, replace the globe on the beam, and determine whether the removal of air has occasioned a loss of weight. If air has weight, what ought to be the effect on the scale-beam if you open the stop-cock and admit air? Try it. Can matter exist in an invisible state? How does nature answer this question in the last experiment?

4. Energy. — Bodies of matter may possess the ability to put other bodies of matter in motion; *e.g.* the bended bow can project an arrow, and the spring of a watch when closely wound can put in motion the machinery of a watch. *Ability to produce motion is called energy.* Nothing but matter possesses energy. *Does air ever possess energy?*

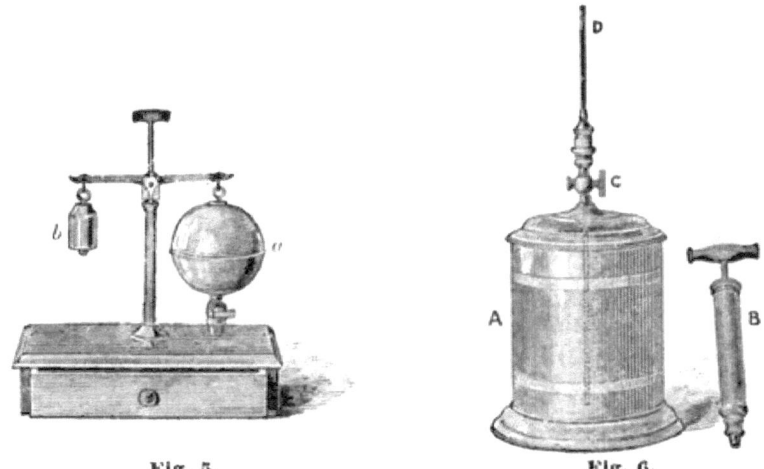

Fig. 5. Fig. 6.

Experiment 4. — Put about one quart of water into vessel A (Fig. 6), called a condensing-chamber. Connect the condensing-syringe B with it, and force a large quantity of air into the portion of the chamber not occupied by water; in other words, fill this portion with condensed air. Close the stop-cock C, and attach the tube D as in the figure. Open the stop-cock, and a continuous stream of water will be projected to a great hight.

Experiment 5. — Remove any water which may remain, and again condense air in the chamber. Connect the chamber by a rubber tube with the nipple *a* of the glass flask (Fig. 7). Place a little water in the neck of the flask, so as to cover the lower orifice of the rotating bulb B. Slowly and carefully open the stop-cock. The escaping air will cause the bulb B to rotate for a long time.

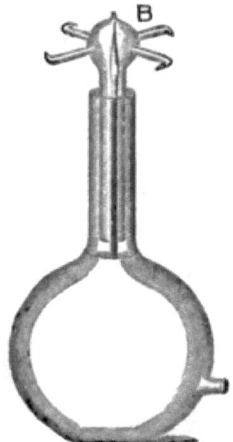

Fig. 7.

You will not attempt to say what matter *is*. This, no one knows. You may, however, give a *provisional* (answering the present needs) definition of matter, *i.e.* draw the limiting line between what is matter and what is not matter.

5. Minuteness of Particles of Matter. — If with a knife-blade you scrape off from a piece of chalk (not from a blackboard crayon, for this is not chalk) a little fine dust, and place it under a microscope, you will probably discover that what seen with the naked eye appear to be extremely small, shapeless particles, are really clusters or heaps of shells and corals more or less broken. Figure 8 represents such a cluster. Each of these shells is susceptible of being broken into thousands of pieces. Reflecting that one of these clus-

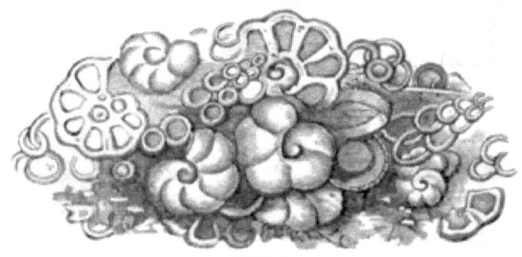

Fig. 8.

ters is so small as to be nearly invisible, you will readily conceive that if one of the shells composing a cluster should be broken into many pieces, and the pieces separated from one another, that they would be invisible to the naked eye. Yet the smallest of the particles into which one of these shells can be broken by pounding or grinding is enormously large in comparison with bodies called *molecules*, which, of course, have never been seen, but in whose existence we have the utmost confidence. (For definition and further discussion of the molecule, see Chemistry, page 4.)[1]

6. Theory of the Constitution of Matter. — For reasons which will appear as our knowledge of matter is extended, physicists have generally adopted the following theory of the constitution of matter: *Every body of matter except the molecule is composed of exceedingly small particles, called molecules. No two molecules of matter in the universe are in permanent contact with each other. Every molecule is in quivering motion, moving back and forth between its neighbors, hitting and rebounding from them.* When we heat a body we simply cause the molecules to move more rapidly through their spaces; so they strike harder blows on their neighbors, and usually push them away a very little; hence, the body expands.

7. Porosity. — If the molecules of a body are never in contact except at the instants of collision, it follows that there are spaces between them. These spaces are called *pores*.

Water absorbs air and is itself absorbed by wood, paper, cloth, etc. It enters the vacant spaces, or pores, between the molecules of these substances. All matter is porous; thus water may be forced through the pores of cast iron; and gold, one of the densest of substances, absorbs liquid mercury.

8. Volume, Mass, and Density. — The quantity of space a body of matter occupies is its *volume*, and is expressed in cubic inches, cubic centimeters, etc. The quantity of matter in a body is its *mass*, and is expressed in pounds,

---

[1] References in this book are made to the Introduction to Chemical Science, by R. P. Williams.

ounces, kilograms, grams, etc. If you cut blocks of wood, potato, cheese, lead, etc., of the same size and weigh them, you will find their weights to be very different. From this you infer that *equal volumes of different substances contain unequal quantities of matter*. Those which contain the greater quantity of matter in the same volume

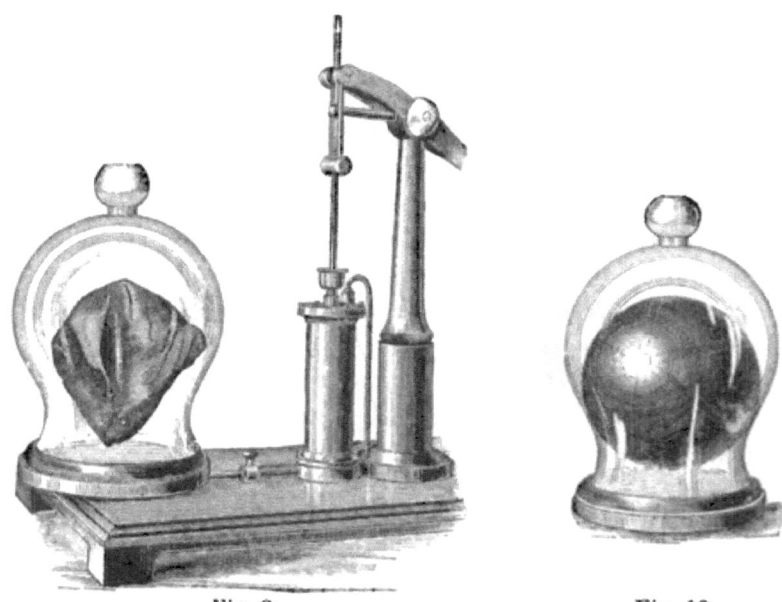

Fig. 9. Fig. 10.

are said to be *denser* than the others. By the *density* of a body is meant *its mass in a unit of volume;* hence it can be expressed only by giving both the units of mass and the unit of volume. For example, the density of cast iron is 4.2 ounces per cubic inch, or 7.2 grams per cubic centimeter; the density of gold is 11 ounces per cubic inch, or 19.4 grams per cubic centimeter. Which of these two metals is the denser?

9. Three States of Matter.

Experiment 6. — Take a thin rubber foot-ball containing very little air, close the orifice of the ball so that air cannot enter or escape, place it under the receiver of an air-pump (Fig. 9), and exhaust the air from the receiver. The air within the ball constantly expands until the ball is completely inflated (Fig. 10).

We recognize three states or conditions of matter, *viz.*, *solid, liquid, and gaseous*, fairly represented by earth, water, and air. Every day observation teaches us that *solids tend to preserve a definite volume and shape; liquids tend to preserve a definite volume only, their shape conforms to that of the containing vessel; gases tend to preserve neither a definite volume nor shape, but to expand indefinitely.*

Liquids and gases in consequence of their manifest tendency to flow are called *fluids*. Even solids possess the property of fluidity to a greater or less extent when under suitable stress. Bodies also exist in intermediate conditions between the solid and liquid, and liquid and gaseous, so that there is no distinct limit between these states, and the distinctions given above are merely conventional (*i.e.* growing out of custom).

Which of the three states any portion of matter assumes depends upon its *temperature and pressure*. Just as at ordinary pressures of the atmosphere water is a solid (*i.e.* ice), a liquid, or a gas (*i.e.* steam), according to its temperature, so any substance may be made to assume any one of these forms unless a change of temperature causes a chemical change, *i.e.* causes it to break up into other substances. For example, wood cannot be melted, because it breaks up into charcoal, steam, etc., before the melting-point is reached. In order that matter may exist in a liquid (and sometimes in a solid) state, a certain definite pressure is required. Ice vaporizes, but does not melt (*i.e.* liquefy) in a space from which the air (and consequently atmospheric pressure) has been removed. Iodine and camphor vaporize, but do not melt unless the pressure is greater than the ordinary atmospheric pressure. Charcoal has been vaporized, but has never been liquefied, undoubtedly because sufficient pressure has never been used.

As regards the temperature at which different substances assume the different states, there is great diversity. Oxygen and nitrogen gases, or air, — which is a mixture of the two, — liquefy and solidify only at extremely low temperatures; and then, only under tremendous pressure. On the other hand, certain substances, as quartz and lime, are liquefied only by the most intense heat generated by an electric current.

Section II.

RELATIVE MOTION AND RELATIVE REST.

10. What constitutes Relative Motion and Relative Rest? — Two boys walk toward each other, or one boy stands, and another boy walks either toward or from him; in either case there is a *relative motion* between them, because *the length of a straight line* (which may be imagined to be stretched) *between them constantly changes.* One boy stands, and another boy walks around him in a circular path; there is a relative motion between them, because *the direction of a straight line between them constantly changes.* There is *relative rest* between two boys while standing, because *a straight line between them changes neither in length nor direction.* Two boys while running are in relative rest so long as neither the distance nor the direction from each other changes.

QUESTIONS.

1. What is wind? Give some evidence that it possesses energy.
2. Give a provisional definition of matter.
3. What is energy?
4. What is an experiment? What is manipulation?
5. What is an air-bubble? What important lesson does a mere bubble teach?

6. What is impenetrability? State several properties that are peculiar to matter.

7. Can water be rendered invisible? How?

8. Under what conditions would a flock of birds over your head be at rest with reference to your body? Would the birds which compose the flock be at rest with reference to one another? An apple rests upon a table; are its molecules at rest?

9. Why do all moving bodies possess energy? Do all molecules possess energy?

10. A span of horses harnessed abreast are drawing a street car on a straight, level road. Is there any relative motion between the two horses? Between the horses and the carriage? Between the team and objects by the wayside? Suppose them to be travelling in a circular path; is there relative motion between the horses?

11. A boat moves away from a wharf at the rate of five miles an hour. A person on the boat's deck walks from the prow toward the stern, at the rate of four miles an hour; what is his rate of motion, *i.e.* his velocity, with reference to the wharf? What is his velocity with reference to the boat?

12. When is there relative motion between two bodies?

Section III.

FORCE.

11. Pushes and Pulls. — We are familiar with the results of muscular force in producing motion. We are also aware that there are forces, or causes of motion, quite independent of man; *e.g.*, the force exerted by wind, running water, and steam. If we observe carefully, we shall find that *all motions are produced by pushes or pulls.* It is evident that *there can be no push or pull except between at least two bodies or two parts of the same body.*

12 MATTER, ENERGY, MOTION, AND FORCE.

Commonly, the bodies between which there is a push or a pull are either in contact, as when we push or pull a table, or the action is accomplished through an intermediate body, as when we draw some object toward us by means of a string, or push an object away with a pole. *Can two bodies push or pull without contact and without any tangible intermediate body;* i.e. *is there ever "action at a distance"?*

Experiment 7. — Fill a large bowl or pail with water to the brim. Place on the surface of the water a half-dozen (or more) floating magnets (pieces of magnetized sewing-needles thrust through thin slices of cork). Hold a bar magnet vertically over the water with one end near, but not touching, the floats; the floats either move toward or away from the magnet. Invert the magnet, and the motions of the floats will be reversed.

Notwithstanding there is no contact or visible connection between the floats and the magnet, the motions furnish conclusive evidence that there are pushes and pulls. The motions are said to be due to *magnetic force*.

Experiment 8. — Suspend two pith balls by silk threads. Rub a large stick of sealing-wax with a dry flannel, and hold it near the balls. The balls move to the wax as if pulled by it, and remain in contact with it for a time. Soon they move away from the wax as if pushed away. Remove the wax; the balls do not hang side by side as at first, but push each other apart (Fig. 11). These motions are said to be due to *electric force*.

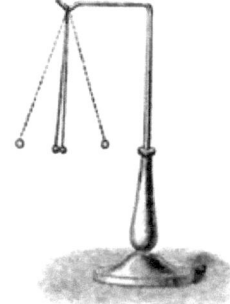
Fig. 11.

12. How Force is Measured. — Pulling and pushing forces may be strong or weak, and are capable of being measured. The common spring balance (Fig. 12) is a very convenient instrument for measuring a pulling force. As usually constructed, the spring balance contains a spiral coil of wire, which is elongated by a pull; and the pulling force is measured by the extent of the elongation. It may be so constructed that an elongated coil may be compressed by a pushing force; and when so constructed it serves to measure a pushing force by the degree of compression. All instruments that measure force, however constructed, are called *dynamometers* (force-measures). Observe that force is measured in pounds; in other words, the unit by which force is measured is called a pound.

Fig. 12.

13. Equilibrium of Forces.

Experiment 9. — Take a block of wood; insert two stout screw-eyes in opposite extremities of the block. Attach a spring balance to each eye. Let two persons pull on the spring balances at the same time, and with equal force, as shown by their indexes, but in opposite directions. The block does not move. One force just neutralizes the other, and the result, so far as the movement of the block, *i.e.* the body acted on, is concerned, is the same as if no force acted on it. When one action, *i.e.* one push or pull, opposes in any degree another action, each is spoken of as a *resistance* to the other. Let f represent the number of pounds of any given force, and let a force acting in any given direction be called *positive*, and indicated by the plus (+) sign, and a force when acting in an opposite direction to a force which we have denominated positive, be called *negative*, and indicated by the minus (−) sign. Then if two forces $+f$ and $-f$ acting on a body at the same point or along the same line are equal, the result is that no change of motion is produced.

Viewed algebraically, $+f-f=0$; or, correctly interpreted, $+f-f \backsimeq$ (is equivalent to) 0, *i.e.* no force. In all such cases there is said to be an *equilibrium of forces*, and the body is said to be in *a state of equilibrium*. If, however, one of the forces is greater than the other, the excess is spoken of as an *unbalanced force*, and its direction is indicated by one or the other sign, as the case may be. Thus, if a force of + 8 pounds act on a body toward the east, and a force of − 10 pounds act on the same body along the same line toward the west, then the unbalanced force is − 2 pounds, *i.e.* the result is the same as if a force of only 2 pounds acted on the body toward the west.

14. Stress, Action, and Reaction; Force Defined. — *An unbalanced force always produces a change of motion.* As there are always two bodies or two parts of a body concerned in every push or pull, there must be two bodies or parts of a body affected by every push or pull. When the effects on both parties to an action are considered without special reference to either alone, the force is frequently called a *stress*. But when we consider the effect on only one of two bodies, we find it convenient, and almost a necessity, to speak of the effect as due to the *action of some other body*, or, still more conveniently, to an *external force*. The body which acts upon another, itself experiences the effect of the *reaction* of the same force.

We may say, provisionally, that *force is that which tends to produce or change motion.* Bringing a body to relative rest is changing its rate of motion and requires force. This definition of force conveys no idea of *what force is;* it merely distinguishes between what is force and what is not force.

QUESTIONS.

1. Give a provisional definition of force. In what two ways is it exerted?

2. How is motion produced? Destroyed? Changed in any way?

3. How many bodies or parts of a body must be concerned in the action of any single force? How many are affected thereby?

4. What effect does an unbalanced force produce on a body?

5. How must the magnitude of two forces compare, and in what directions must they act with reference to each other, that they may be in equilibrium?

6. When is a body in equilibrium?

7. In what units is force estimated? In what units is mass estimated? What force is required to support 10 pounds of sugar? What is the common way of judging of the mass of a body?

8. Why will not a force of 10 pounds raise 10 pounds of sugar? If the force produces no change of motion, how can it consistently be called a force?

9. A bullet is flying unimpeded through space; does it possess energy? Is it (disregarding the force of gravity) exerting force? Would it exert force if it should encounter some other body? Which produces motion, energy or force? Which denotes ability to produce motion?

Section IV.

ATTRACTION OF GRAVITATION.

15. Gravitation is Universal. — An unsupported body falls to the earth. This is evidence of an action or stress between the earth and the body. It has been ascertained by careful observation that when a ball is suspended by a long string by the side of a mountain, the string is not quite vertical, but is deflected toward the mountain in consequence of an attraction between the mountain and the ball. That there is an attraction between the sun and the earth, and the earth and the moon, is shown, as we shall see further on, by their curvilinear motions. Tides and tidal currents on the earth are due to the attraction of the sun and the moon.

This attraction is called *gravitation*; the force is called *gravity*. When bodies under its influence tend to approach one another, they are said to *gravitate*. Since this attraction ever exists between all bodies, at all distances, it is called *universal gravitation*.

16. Law of Universal Gravitation. — Methods too difficult for us to comprehend at present have established the fact that the strength of the attraction between any two bodies depends upon two things; *viz.*, their *masses*, and the *distance* between certain points within the bodies (to be explained hereafter), called their centers of gravity. The following law is found everywhere to exist: —

The attraction between every two bodies of matter in the universe varies directly as the product of their masses, and inversely as the square of the distance between their centers of gravity. Representing the masses of two bodies by m and m', the distance by d, and the attraction by g, this relation is expressed mathematically, thus: $g \propto$ (varies as) $\frac{mm'}{d^2}$. For example, if the mass of either body is doubled, the product (mm') of the masses is doubled, and consequently the attraction is doubled. If the distance between their centers of gravity is doubled, then $\left(\frac{1^2}{2^2} = \frac{1}{4}\right)$ the attraction becomes one-fourth as great.

The mass of the moon is very much less than that of the earth; hence the force of gravity at the surface of the former is much less than at the surface of the latter. A person who could leap a fence three feet high on the earth, could, by the exertion of the same muscular energy, leap a fence 18 feet high on the moon. A boy might throw a stone a greater distance on the moon than a rifle can project a bullet on the earth. The masses of Jupiter and Saturn, being so much greater than that of the earth, the corresponding greater attraction which they would exert would so impede locomotion that a person would be able only to crawl along as though his feet were weighted with lead.

17. Weight. — We say that all matter has weight, meaning that there is an attraction between the earth and all kinds of matter. We say that the weight of a certain body is ten pounds, meaning that this is the measure of the force of attraction between this body and the earth.

From the law of gravitation we infer that *at equal distances from the earth's center of gravity the weight of bodies varies as their masses.* Hence, when we weigh a body we measure at the same time both the force with which the earth attracts it and its mass; and both quantities are commonly expressed in units of the same name. The expression *four pounds of tea* conveys the twofold idea that the quantity of tea is four pounds, and that the force with which the earth attracts the tea is four pounds.

Again, we infer from the law of gravitation (1) that *a body weighs more at a given point on the surface of the earth than at any point above this point.*

(2) That inasmuch as some points on the earth's surface are nearer its center of gravity than others, *the same body will not have the same weight at all points on the earth's surface.* A given body stretches a spring balance less as it is carried from either pole toward the equator. The loss of weight due to the increase of distance from the center of the earth is $\frac{1}{568}$ of its weight at the poles.

18. Point of Maximum Weight. — There is no definite law which determines the change in the weight of a body when carried below the surface of the earth. Observation has shown that at first a body increases in weight slowly, in consequence of its approach to the earth's center of gravity. But at some undetermined depth, in consequence of an increase of density of the earth toward its center, the increase of weight must cease; and at this point, consequently, a body has its *maximum weight.* From this point onward to the center of gravity of the earth, a body will lose in weight as much as it would if it were being transferred to smaller and smaller earths.

QUESTIONS.

1. If the earth's mass were doubled without any change of volume, how would it affect your weight?

2. On what principle do you determine that the mass of one body is ten times the mass of another body?

3. How many times must you increase the distance between the centers of two bodies that their attraction may become one-fourth?

4. If a body on the surface of the earth is 4,000 miles from the center of gravity of the earth, and weighs at this place 100 pounds, what would the same body weigh if it were taken 4,000 miles above the earth's surface?

5. The masses of the planets Mercury, Venus, Earth, and Mars are respectively very nearly as 7, 79, 100, and 12; assuming that the distance between the centers of the first two is the same as the distance between the centers of the last two, how would the attraction between the first two compare with the attraction between the last two?

6. What would be the answer to the last question if the distance between the centers of the first two were four times the distance between the centers of the last two?

7. Would the weight of a soldier's knapsack be sensibly less if it were carried on the top of his rifle?

Section V.

MOLECULAR FORCES.

19. Molecular Distinguished from Molar Forces; Repellent Force. — Thus far we have considered only the effects of the action of bodies of sensible (perceived by the senses) size and at sensible distances. Have we any evidence that the molecules which compose these bodies act upon one another in a similar manner?

If you attempt to break a rod of wood or iron, or stretch a piece of rubber, you realize that there is a force resisting you. You reason that if the supposition be true, that the grains or molecules that compose these bodies do not touch one another, then there must be a powerful *attractive force* between the molecules, to prevent their separation. After stretching the rubber, let go one end; it springs back to its original form. What is the cause? The volume of most bodies is diminished by compression; when the pressure is removed, they recover to a greater or less extent their previous volume. What is the cause?

Every body of matter, with the possible exception of the molecule, whether solid, liquid, or gaseous, may be forced into a smaller volume by pressure; in other words, *matter is compressible*. When pressure is removed, the body expands into nearly or quite its original volume. This shows two things: first, that *the matter of which a body is formed does not really fill all the space which the body appears to occupy;* and, second, that *in the body is a force which resists outward pressure tending to compress it, and expands the body to its original volume when pressure is removed*. This is, of course, a *repellent force*, and is exerted among molecules, tending to push them farther apart.

For convenience, we call bodies of appreciable size *molar* (massive) in distinction from molecules (bodies of very small mass). Action between molar bodies, usually at sensible distances, is called *molar force;* action between molecules, always at insensible distances, is called *molecular force*.

20. Cohesion, Tenacity. — That attraction which holds the molecules of the same substance together, so as to form larger bodies, is called *cohesion*. It is the attraction that resists a force tending to break or crush a body. The *tenacity* of solids and liquids, *i.e.* the resistance which they offer to being pulled apart, is due to this attraction. It is greatest in solids, usually less in liquids, and entirely wanting in gases. It acts only at insensible distances, and is strictly molecular. When cohesion is overcome, it is usually difficult to force the molecules near enough to one another for this attraction to become effective again. Broken pieces of glass and crockery cannot be so nicely readjusted that they will hold together. Yet two polished surfaces of glass or metal, placed in contact, will cohere quite strongly. Or if the glass is heated till it is soft, or in a semi-fluid condition, then, by pressure, the molecules at the two surfaces will flow around one another, pack themselves closely together, and the two bodies will become firmly united. This process is called *welding*. In this manner iron is welded.

Cohesive force varies greatly both in intensity and its behavior in different substances, and even in the same substances under different circumstances. Modifications of this force give rise to certain *conditions of matter* designated as *crystalline or amorphous, hard or soft, flexible or rigid, elastic, viscous, malleable, ductile, tenacious,* etc.

21. Crystallization.

Fig. 13.

Experiment 10. — Pulverize about three ounces of alum. Take about a teacupful of boiling hot water in a beaker, and sift into it the powdered alum, stirring with a glass rod as long as the alum will dissolve readily. Then suspend in the liquid to a little depth one or more threads from a splinter of wood laid across the top of a beaker (Fig. 13). Place the whole where it will not be disturbed, and allow it to cool slowly. It is well to allow it to stand for a day or more.

MOLECULAR FORCES. 21

Beautiful transparent bodies of regular shape are formed on the bottom and sides of the beaker and probably on the thread. They are called *crystals*, and the process by which they are formed is called *crystallization*.

Observe that the crystals formed on the thread in mid-liquid are much more regular in shape than those formed on the surface of the glass. The latter are flattened, and are said to be *tabular*.

In a similar manner, obtain crystals of bichromate of potash, blue vitriol, copperas, etc. Make up a cabinet of crystals, preserving them in small, closely stopped glass bottles.

Experiment 11. — Thoroughly clean a piece of window glass, by breathing upon it, and then rubbing it with a piece of newspaper.

Fig. 14.

Warm the glass over an alcohol or Bunsen flame, and pour upon the glass a strong solution of sal ammoniac, or saltpetre. Allow the liquid to drain off, and hold the wet glass up to the sunlight, or view it through a magnifying glass, and watch the growth of the crystals.

Experiment 12. — Examine with a magnifying glass the surface fracture of a freshly broken piece of sugar loaf, and observe, if any, small, smooth, glistening planes thus exposed.

These planes are surfaces of small, imperfectly formed crystals closely packed together, similar to the imperfect crystals of alum, etc., formed on the sides of the beaker. Such bodies are said to have a *crystalline fracture*, and the body itself is said to be crystalline in distinction from *amorphous* matter like glass, glue, etc., which furnish no evidence of crystalline structure.

Very interesting illustrations of crystallization are those delicate lace-like figures which follow the touch of frost on the window-pane. Figure 14 represents a few of more than a thousand forms of snowflakes that have been discovered, resulting from a variety of arrangement of the water molecules.

Snow crystals are formed during free suspension of moisture in the air and without interference from contact with any solid; hence their perfection of growth. If you gather snowflakes, as they fall, on cold, yellow glass and examine them under a magnifying glass, you will find that all crystals have a *primary type of six rays, and hexagonal outline*. Professor Tyndall has succeeded in so unravelling lake ice as to show what he calls "liquid flowers" in a block of ice, thus proving that ice is crystalline, or composed of a compact mass of crystals. (Read Tyndall's "Forms of Water.")

Nature teems with crystals. Nearly every kind of matter, in passing from the liquid state (whether molten or in solution) to the solid state, tends to assume symmetrical forms. *Crystallization is the rule; amorphism, the exception*. You can scarcely pick up a stone and break it without finding the same crystalline fracture.

The massive pillars of basaltic rock found in certain localities, for example, in Fingal's Cave (Fig. 15), might in its broadest sense be regarded as forms of crystallization, inasmuch as they are the result of natural causes. These hexagonal columns, however, probably resulted from great lateral pressure, exerted while cooling, upon molten matter thrown up ages ago by submarine volcanoes.

This tendency of the molecules of matter to arrange themselves in definite ways during solidification is attended usually with a change of volume. The molecular force exerted at such a time is sometimes enormous, so as to burst the strongest vessels. Hence our service pipes are burst when water is allowed to crystallize (freeze) in them.

22. Hardness.

Experiment 13. — Get specimens of the following substances: talc, chalk, glass, quartz, iron, silver, lead, copper, rock-salt, and marble. Ascertain which of them will scratch glass, and which are scratched by glass. Which is the softest metal that you have tried? The hardest? Name some metal that you can scratch with a finger-nail. See if you can scratch a piece of copper with a piece of lead, and *vice versa*. Which is softer, iron or lead? Which is the denser metal? Does hardness depend upon density? What force must be overcome in order to scratch a substance?

Fig. 15.

To enable us to express degrees of hardness, the following table of reference is generally adopted: —

MOHR'S SCALE OF HARDNESS.

1. Talc.
2. Gypsum (or Rock-Salt).
3. Calcite.
4. Fluor-Spar.
5. Apatite.
6. Orthoclase (Feldspar).
7. Quartz.
8. Topaz.
9. Corundum.
10. Diamond.

By comparing a given substance with the substances in the table, its degree of hardness can be expressed approximately by one of the numbers used in the table. If the hardness of a substance is indicated by the number 4, what would you understand by it?

23. Hardening and Annealing; Flexibility.

Experiment 14. — Get pieces of wire, each ten inches long, of the following metals: steel, iron, spring brass, hard copper, German silver, platina, and phosphor-bronze. Place each in an alcohol or Bunsen flame, and heat the wire near one end to a bright red glow, and then thrust the heated part into cold water, and suddenly cool it. See whether the part thus treated bends more or less readily than the part which has not suffered the sudden change. When a body is easily bent, *i.e.* its cohesive force admits of a hinge-like movement among its molecules without permanent separation, it is said to be *flexible.* See whether the part treated has been hardened or softened by the treatment. The process of rendering flexible and softening is called *annealing.*

Next heat the opposite ends of the wires as before, and slowly (10 to 15 minutes) withdraw the wires from the flame by gradually raising them above the flame, in order that the fall of temperature may be very gradual. Ascertain as before the effect of this treatment on the flexibility and hardness of each. Classify the substances as *annealed by sudden cooling*, and *annealed by slow cooling.*

24. Elasticity.

Experiment 15. — Obtain thin strips of as many of the following substances as practicable: rubber, different kinds of wood, ivory, whalebone, steel, spring brass and soft brass, copper, iron, zinc, and lead.

Bend each one of the above strips. Note which completely unbends when the force is removed. Arrange the names of these substances in the order of the rapidity and completeness with which they unbend.

The property which matter possesses of recovering its former shape and volume, after having yielded to some force, is called *elasticity.*

25. Viscosity.

Experiment 16. — Support in a horizontal position, at one of its extremities, a stick of sealing-wax, and suspend from its free extremity an ounce weight, and let it remain in this condition several days, or perhaps weeks. At the end of the time the stick will be found permanently bent. Had an attempt been made to bend the stick quickly, it would have been found quite brittle. A body which, subjected to a stress for a considerable time, suffers a permanent change in form is said to be *viscous.* Hardness is not opposed to viscosity. A lump of pitch may be quite hard, and yet in the course of time it will flatten itself out by its own weight, and flow down hill like a stream of syrup.

Sealing-wax and pitch may be regarded as *fluids* whose flow is extremely slow; *i.e.* their viscosity or resistance to flow is very great. Liquids like molasses and honey are said to be viscous, in distinction from limpid liquids like water and alcohol.

26. Malleability and Ductility.

Experiment 17. — Place a piece of lead on an anvil, or other flat bar of surface, and hammer it. It spreads out under the hammer into sheets, without being broken, though it is evident that the molecules have moved about among one another, and assumed entirely different relative positions. Heat a piece of soft glass tube in a gas-flame, and, although the glass does not become a liquid, it behaves very much like a liquid, and can be drawn out into very fine threads.

When a solid possesses sufficient *fluidity* to admit of being drawn out into threads, it is said to be *ductile*. When it will admit of being hammered or rolled into sheets, it is said to be *malleable*.

Platinum and gold are the most malleable and ductile metals. They can be drawn into wire finer than a spider's thread, or so as to require very keen vision to see it. Gold can be hammered into leaves $\frac{1}{300000}$ of an inch thick. Some metals, like iron, are more malleable and ductile at a red heat; others, like copper, at an ordinary temperature.

It is remarkable that the tenacity of most metals is increased by being drawn out into wires. It would seem that, in the new arrangement which the molecules assume, the cohesive force is stronger than in the old. Hence cables made of iron wire twisted together, so as to form an iron rope, are stronger than iron chains of equal weight and length, and are much used instead of chains where great strength is required.

27. Adhesion.

— If you touch with your finger a piece of gold-leaf, it will stick to your finger; it will not drop off, it cannot be shaken off; and an attempt to pull it off increases the difficulty. Dust and dirt stick to clothing. Thrust your hand into water, and it comes out wet. We could not pick up anything, or hold anything in our hands, were it not that these things stick to the hands.

Every minute's experience teaches us that not only is there an attractive force between molecules of the same kind of matter, but there is also an attractive force between molecules of unlike matter. That force which causes unlike substances to cling together is called *adhesion*. It is probable that *there is some adhesion between all substances when brought in contact.* Glass is wet by water, but is not wet by mercury. *If a liquid adheres to a solid more firmly than the molecules of the liquid cohere, then will the solid be wet by the liquid.* If a solid is not wet by a liquid, it is not because adhesion is wanting, but because cohesion in the liquid is stronger.

28. Tension. — When a rubber band or cord is pulled or stretched, it is said to be in a state of *tension* (*i.e.* of being stretched). The amount of tension in a string supporting a stone is the weight of the stone. A rubber balloon inflated with compressed air is in a state of tension; the air within is in a state of unusual compression. Gases are ever in a state of compression, since they ever tend to expand without limit.

29. Surface Tension. — The molecular forces of cohesion and adhesion give rise to a remarkable series of phenomena, especially obvious in liquids, known as phenomena of *surface tension*. The general law governing all of this class of phenomena is that *the surfaces of all bodies tend to contract indefinitely.* Since solids are those bodies which tend to resist any force tending to alter their shape, and gases have no surfaces of their own, it is obvious why liquids show the effects of such a force most readily. The tendency of a surface of liquid to contract is illustrated in an imperfect manner by a stretched sheet of rubber; the latter, however, has a constantly decreasing force of contraction as it approaches its original dimensions, and it may have a contractile force in only one direction, while a surface sheet of liquid always tends to contract with the same force independently of its size, and it is exerted alike in all directions.

As a consequence of this, *every body of liquid tends to assume the spherical form*, since the sphere has less surface than any other form having equal volume. In large bodies the distorting forces due to gravity are generally sufficient to disguise the effect; but in small bodies, as in drops of water or mercury, it is apparent. Again, if the distorting effect of weight is eliminated in any way, as by immersing a quantity of oil in a mixture of water and alcohol of its own density, or by replacing the central portion of the body by a fluid much lighter than its own kind, as in the case of a soap-bubble, the sphere is the resulting form.

Experiment 18. — Form a soap-bubble at the orifice of the bowl of a tobacco pipe, and then, removing the mouth from the pipe, observe that tension of the two surfaces (exterior and interior) of the bubble drives out the air from the interior and finally the bubble contracts to a flat sheet.

30. Capillary Phenomena. — As a result of molecular action it is found that the surface of a given liquid will always meet a given solid at a definite angle; thus the surface separating water and air always meets clean glass at a very small angle (Fig. 15a); that separating mercury and air meets glass at an angle of about 135°. If clean silver is substituted for glass, the first angle becomes large, not far from 90°, while the second would be reduced to zero; in other words, the mercury creeps along the surface of silver, its own air-exposed surface being parallel with that of the silver.

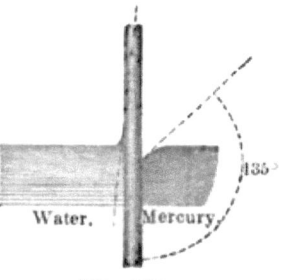

Fig. 15a.

From this it follows, that if a glass tube be dipped into water, the surface tension will cause the liquid to *rise in the bore of the tube above its level outside;* while, on the contrary, if the tube be dipped into mercury, there will result a *depression.* These phenomena are known respectively as *capillary ascension* and *capillary depression.*

If the bore of the tube is reduced one-half in diameter, the lifting force is reduced one-half, but the cross-section will be reduced to one-fourth; hence in order that the weight of the liquid lifted may be one-half, it must rise twice as high as before. Thus we have the law that the *ascension (or depression) of a liquid in a capillary tube is inversely proportional to the diameter of the bore.*

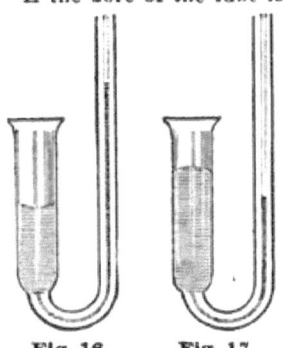

Fig. 16.   Fig. 17.

Experiment 19. — Take a clean glass tube of capillary (*i.e.* small, hair-like) bore, and thrust one end to a depth of about a quarter of an inch in water. Does the water ascend or descend a little way in the tube? What is the shape of the surface of the water in the bore of the tube? Is the edge of the water next the tube on the outside turned up or down?

Repeat the experiment with tubes having bores of different size. Do you notice any difference in the phenomena in the different tubes? If so, in which are the phenomena most striking?

Repeat all the above experiments, and answer all the above questions, using mercury instead of water.

Experiment 20. — Pour a little water into a U-shaped tube (Fig. 16). one of whose arms has a capillary bore; how does the water behave in the capillary tube? Pour a little mercury into another similar tube (Fig. 17);

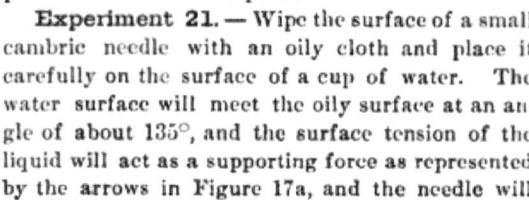

Fig. 17a.

how does the mercury behave? Describe the upper surfaces of both liquids.

Experiment 21. — Wipe the surface of a small cambric needle with an oily cloth and place it carefully on the surface of a cup of water. The water surface will meet the oily surface at an angle of about 135°, and the surface tension of the liquid will act as a supporting force as represented by the arrows in Figure 17a, and the needle will float in a trough-shaped depression in the liquid surface.

QUESTIONS.

1. Why are pens made of steel? What moves the machinery of a watch? What is the cause of the softness of a hair mattress or feather-bed? On what does the entire virtue of a spring balance depend?

2. What name would you give to the attraction which causes your hands to be wet by a liquid? Is adhesion a molar or a molecular force?

3. The tension of a violin string is 2 pounds; what is meant by this statement?

4. Why are liquid drops round? Why are bubbles round?

CHAPTER II.

DYNAMICS[1] *OF FLUIDS.*

Section I.

PRESSURE IN FLUIDS.

31. Cause of Pressure. — We live above a watery ocean and at the bottom of an exceedingly rare and elastic aerial ocean, called the *atmosphere*, extending with a diminishing density to an undetermined distance into space. Every molecule, in both the gaseous and liquid oceans, is drawn toward the earth's center by gravity. This gives to both fluids a downward pressure upon everything on which they rest.

The gravitating action of liquids is everywhere apparent, as in the fall of drops of rain, the descent of mountain streams, and the weight of water in a bucket. But to perceive that air exerts a downward pressure requires special manipulation. If we lower a pail into a well, it fills with water, but we do not perceive that it becomes heavier thereby; the weight of the water in the pail is not felt. But when

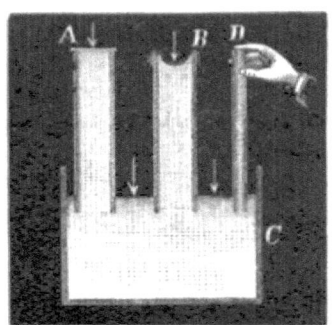

Fig. 18.

we raise a pailful out of the water, it suddenly appears

---

[1] Dynamics is the science which investigates the action of force.

heavy. If we could raise a pailful of air out of the ocean of air, might not the weight of the air become perceptible? If we dive to the bottom of a pond of water, we do not feel the weight of the pond resting upon us. We do not feel the weight of the atmospheric ocean resting upon us; but we should remember that our situation with reference to the air is like that of a diver with reference to water.

32. Gravity causes Pressure in All Directions.

Experiment 22. — Fill two glass jars (Fig. 18) with water, A hav-

Fig. 19.

ing a glass bottom, B a bottom provided by tying a piece of sheet-rubber tightly over the rim. Invert both in a larger vessel of water, C. The water in A does not feel the downward pressure of the air directly above it, the pressure being sustained by the rigid glass bottom. But it indirectly feels the pressure of the air on the surface of the water in the open vessel, and it is this pressure that sustains the water in the jar. But the rubber bottom of the jar B yields somewhat to the downward pressure of the air, and is forced inward.

Experiment 23. — Fill a glass tube, D, with water, keeping one end in the vessel of water, and a finger tightly closing the upper end. Why does not the water in the tube fall? Remove your finger from the closed end. Why does the water fall?

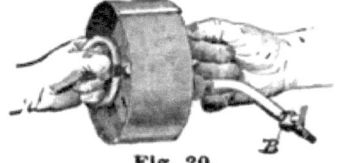

Fig. 20.

Experiment 24. — Fill (or partly fill) a tumbler with water, cover the top closely with a card or writing-paper, hold the paper in place with the palm of the hand, and quickly invert the tumbler (Fig. 19). Why does not the water fall out?

Experiment 25. — Force the piston A (Fig. 20) of the seven-in-one apparatus (so called from the number of experiments that may be performed with one piece of apparatus) quite to the closed end of the hollow cylinder, and close the stop-cock B. Try to pull the piston out again. Why do you not succeed? Hold the apparatus in various positions, so that the atmosphere may press down, laterally, and up against the piston. Do you discover any difference in the pressure which it receives from different directions?

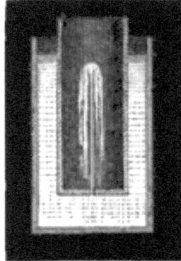
Fig. 21.

Experiment 26. — Force a tin pail (Fig. 21), having a hole in its bottom, as far as possible into water, without allowing water to enter at the top. A stream of water spurts through the hole. Why? Why does it require so much effort to force the pail down into the water?

33. Comparison of Pressure at the Same Depth in Different Directions.

Experiment 27. — Take a glass tube about 30 inches long and one-fourth inch bore, and bend it into the shape of A (Fig. 22). Also prepare tubes like B and C. Let the bend a be about half full of water. Slowly lower the end n into a tumbler filled with water. The water presses up against the air in the tube, and the air transmits the pressure to the liquid in the bend. How is the pressure affected by depth? Does it increase *as* the depth?

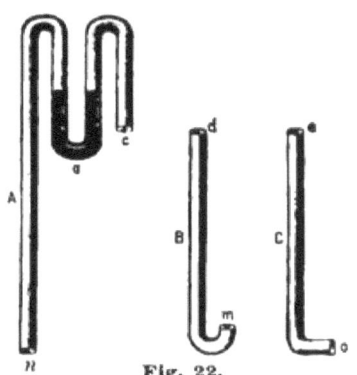
Fig. 22.

Experiment 28. — Connect c with d by means of a rubber tube, and lower the extremity m into the tumbler of water. As the tube is turned up, the water must now press *down* the tube against the air. Does the *downward* pressure increase *as* the depth?

Experiment 29. — Connect e with c, and lower o into the water. The water now presses laterally (sidewise) against the air. Does the *lateral* pressure increase *as* the depth?

Experiment 30. — Fill two tumblers with water, and lower n into one and o into the other, keeping both extremities at the same depth in the liquids. How is the liquid in the bend a affected? How do

DYNAMICS OF FLUIDS.

Fig. 23.

the upward and lateral pressures at the same depth compare?

Experiment 31. — Once more connect c with d, and lower n and m to the same depth into the water in the two tumblers. How do the upward and downward pressures at the same depth compare? *At the same depth is pressure equal in all directions?*

Experiment 32. — Connect the two brass tubes at the extremities F and G (Fig. 23). Fill the cup of the (eight-in-one) apparatus with water, and remove the caps A, B, C, and D from the branch tubes, so as to permit water to escape from the orifices at their ends. Does the water issuing from these orifices show a lateral pressure? What difference do you observe in the flow of water from the different orifices? How do you account for it?

The results of experiments thus far show that *at every point in a body of fluid gravity causes pressure to be exerted equally in all directions, and that in liquids the pressure increases as the depth increases.*

Section II.

MEASUREMENT OF ATMOSPHERIC PRESSURE, BAROMETERS.

34. How Atmospheric Pressure is Measured.

Experiment 33 (preliminary). — Take a U-shaped glass tube (Fig. 24), half fill it with water, close one end with a thumb, and tilt the tube so that the water will run into the closed arm and fill it; then restore it to its original vertical position. Why does not the water settle to the same level in both arms?

Fig. 24.

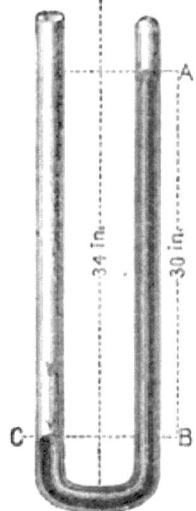

Fig. 25.

Figure 25 represents a U-shaped glass tube closed at one end, 34 inches in hight, and with a bore of 1 square inch section. The closed arm having been filled with mercury, the tube is placed with its open end upward, as in the cut. The mercury in the closed arm sinks about 2 inches to A, and rises 2 inches in the open arm to C; but the surface A is 30 inches higher than the surface C. This can be accounted for only by the atmospheric pressure. The column of mercury BA, containing 30 cubic inches, is an exact counterpoise for a column of air of the same diameter extending from C to the upper limit of the atmospheric ocean, — an unknown hight.

The weight of the 30 cubic inches of mercury in the column BA is about 15 pounds. Hence the weight of a column of air of 1 square-inch section, extending from the surface of the sea to the upper limit of the atmosphere, is about 15 pounds. But in fluids gravity causes equal pressure in all directions. Hence, *at the level of the sea, all bodies are pressed upon in all directions by the atmosphere, with a force of about 15 pounds per square inch, or about one ton per square foot.*

A pressure of 15 pounds per square inch is quite generally adopted as a unit of gaseous pressure, and is called an *atmosphere*.

Fig. 26.

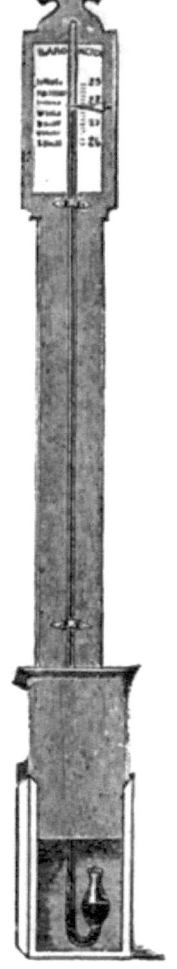

Fig. 27.

35. Barometer. — The hight of the column of mercury supported by atmospheric pressure is quite independent, however, of the area of the surface of the mercury pressed upon; hence the apparatus is more conveniently constructed in the form represented in Figure 26.

A straight tube about 34 inches long is closed at one end and filled with mercury. A finger tightly closing the open end, the tube is inverted, and this end is inserted in a vessel of mercury and the finger is withdrawn, when the mercury sinks until there is equilibrium between the downward pressure of the mercurial column AB and the pressure of the atmosphere. An apparatus designed to measure atmospheric pressure is called a *barometer* (pressure-measurer). A common form of barometer is represented in Figure 27. Beside the tube and near its top is a scale graduated in inches or centimeters, indicating the hight of the mercurial column. For ordinary purposes this scale needs to have only a range of three or four inches, so as to include the maximum fluctuations of the column.

The hight of the barometric column is subject to fluctuations; this shows that the atmospheric pressure is subject to variations. The barometer is always a faithful monitor of all changes in atmospheric pressure. It is also serviceable as a weather indicator. It does not indicate weather that is present, but foretells coming weather. Not that any particular point at which mercury may stand foretells any particular kind of weather, but *any sudden change in the barometer indicates a change in the weather.* A rapid fall of mercury generally forebodes a storm, while a rising column indicates clearing weather.

36. Aneroid Barometer. — The *aneroid* (without moisture) barometer employs no liquid. It contains a cylindrical box, D (Fig. 28), having a very flexible top. The air is partially exhausted from within the box. The varying atmospheric pressure causes this top to rise and sink much like the chest of man in breathing. Slight movements of this kind are communicated by means of multiplying-apparatus (apparatus by means of which a small movement of one part is magnified into a large movement of another part) to the index needle A. The dial is graduated to correspond with a mercurial barometer. The observer turns the button C and brings the brass needle B over the black needle A, and at his next observation any departure of the latter from the former will show precisely the change which has occurred between the observations.

The aneroid can be made more sensitive (*i.e.* so as to show smaller changes of atmospheric pressure) than the mercurial barometer. If a barometer is carried up a mountain, it is found that the mercury constantly falls as the ascent increases. Roughly speaking, the barometer falls one inch for every 900 feet of ascent. Really, in consequence of the rapid increase of the rarity of the air, the rate of fall diminishes as you ascend. It is obvious that the barometer will serve to measure approximately the hights of mountains.

Fig. 28.

If a mercurial barometer stand at 760 mm on the floor, the same barometer on the top of a table 1m high should stand at a hight of 759.91mm, a change scarcely perceptible. The aneroid is, however, sometimes made so sensitive that the change of pressure experienced in this short distance is rendered quite perceptible.

The shading in Figure 29 is intended to indicate roughly the variation in the density of the air at different elevations above sea-level. The figures in the left margin show the hight in miles; those in the first column on the right, the corresponding average hight of the mercurial

BAROMETERS. 37

column in inches; and those in the extreme right, the density of the air compared with its density at sea-level. The average hight of the mercurial column at sea-level is about 30 inches (76^{cm}).

If an opening could be made in the earth, 35 miles in depth below the sea-level, it is calculated that the density of the air at the bottom would be 1,000 times that at sea-level, so that water would float in it. Air has been compressed to this density.

To what hight the atmosphere extends is unknown. It is variously estimated at from 50 to 200 miles. If the aerial ocean were of uniform density, and of the same density that it is at the sea-level, its depth would be a little short of five miles. Certain peaks of the Himalayas would rise above it.

Fig. 29.

Section III.

COMPRESSIBILITY AND ELASTICITY OF GASES. — BOYLE'S LAW.

37. Compressibility of Gases. — The increase of pressure attending the increase in depth, in both liquids and gases, is readily explained by the fact that the lower layers of fluids sustain the weight of all the layers above. Consequently, if the body of fluid is of uniform density, as is very nearly the case in liquids, the pressure will increase in nearly the same ratio as the depth increases. But the aerial ocean is far from being of uniform density, in consequence of the extreme *compressibility* of gaseous matter. The contrast between water and air, in this respect, may be seen in the fact that water subjected to a pressure of one atmosphere contracts 0.0000457 of its volume; under the same circumstances, air contracts one-half. For most practical purposes, we may regard the density of water at all depths as uniform, while it is far otherwise in large masses of gases.

38. Elasticity of Gases. — Closely allied to compressibility is the *elasticity* of gases, or their power to recover their former volume after compression. *The elasticity of all fluids is perfect.* By this is meant, that the force exerted in expansion is equal to the force used in compression; and that, however much a fluid is compressed, it will always completely regain its former bulk when the pressure is removed. Hence the barometer which measures the compressing force of the atmosphere also measures at the same time the elastic force (*i.e.* the

COMPRESSIBILITY AND ELASTICITY OF GASES. 39

tension or expansive force) of the air. Liquids are perfectly elastic; but, inasmuch as they are perceptibly compressed only under tremendous pressure, they are regarded as practically incompressible, and so it is rarely necessary to consider their elasticity. It has already been stated that matter in a gaseous state expands indefinitely unless restrained by external force. The atmosphere is confined to the earth by the force of gravity.

Experiment 34. — Force the piston of the seven-in-one apparatus two-thirds the way into the cylinder, and close the aperture. Support the apparatus on blocks, with the piston upwards, remove the handle, and place a weight on the piston, and place the whole under the receiver of an air-pump. Exhaust the air from the receiver; the outside pressure of the air being partially removed, the unbalanced force (*i.e.* the tension) of the air enclosed within the cylinder will cause the piston to rise, and raise the weight.

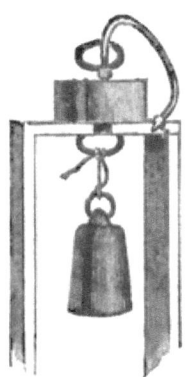

Fig. 30.

Experiment 35. — Arrange the same apparatus as in Figure 30. Attach a small rubber tube to the short tube, and suck as much air out of the cylinder as possible. The air within, being rarefied, loses its tension, and the unbalanced outside pressure forces the piston into the cylinder, raising the weight. A very much heavier weight may be raised if the rubber tube connects the apparatus with an air-pump.

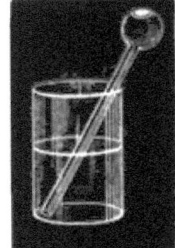

Fig. 31.

Experiment 36. — Take a glass tube (Fig. 31) having a bulb blown at one end. Nearly fill it with water, so that when inverted there will be only a bubble of air in the bulb. Insert the open end in a glass of water, place under a receiver, and exhaust. Nearly all the water will leave the bulb and tube. Why? What will happen when air is admitted to the receiver?

39. Boyle's or Mariotte's Law.

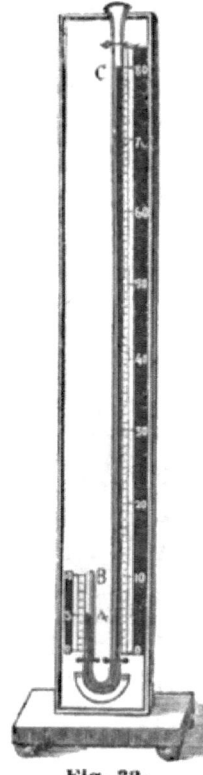

Fig. 32.

Experiment 37. — Take a bent glass tube (Fig. 32), the short arm being closed, and the long arm, which should be at least 34 inches (85^{cm}) long, being open at the top. Pour mercury into the tube till the surfaces in the two arms stand at zero. Now the surface in the long arm supports the weight of an atmosphere. Therefore the tension of the air enclosed in the short arm, which exactly balances it, must be about 15 pounds to the square inch. Next pour mercury into the long arm till the surface in the short arm reaches 5, or till the volume of air enclosed is reduced one-half, when it will be found that the hight of the column AC is just equal to the hight of the barometric column at the time the experiment is performed. It now appears that the tension of the air in AB balances the atmospheric pressure, *plus* a column of mercury AC, which is equal to another atmosphere; ∴ the tension of the air in AB = two atmospheres. But the air has been compressed into half the space it formerly occupied, and is, consequently, twice as dense. If the length and strength of the tube would admit of a column of mercury above the surface in the short arm equal to twice AC, the air would be compressed into one-third its original bulk; and, inasmuch as it would balance a pressure of three atmospheres, its tension would be increased threefold.

From this experiment we learn that, at twice the pressure there is half the volume, while the density and elastic force are doubled. Hence the law:—

The volume of a body of gas at a constant temperature varies inversely as the pressure, density, and elastic force.

For many years after the announcement of this law it was believed to be rigorously correct for all gases, but more recently, more precise experiments have shown that it is approximately but not rigidly true for any gas, that the departure from the law differs with different gases, and that *each gas possesses a special law of compressibility.*

Section IV.

INSTRUMENTS USED FOR RAREFYING AND CONDENSING AIR.

40. The Air-Pump. — The air-pump, as its name implies, is used to withdraw air from a closed vessel. Figure 33 will serve to illustrate its operation. R is a glass *receiver* from which air is to be exhausted. B is a hollow cylinder of brass, called the *pump-barrel*. The plug P, called a *piston*, is fitted to the interior of the barrel, and can be moved up and down by the handle H; *s* and *t* are valves.

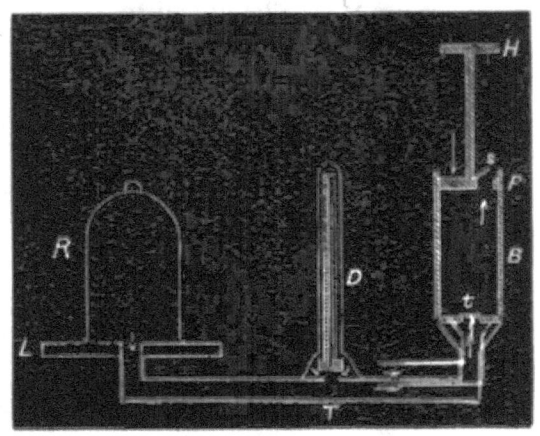

Fig. 33.

A valve acts on the principle of a door intended to open or close a passage. If you walk against a door on one side, it opens and allows you to pass; but if you walk against it on the other side, it closes the passage, and stops your progress. Suppose the piston to be in the act of descending; the compression of the air in B closes the valve *t*, and opens the valve *s*, and the enclosed air escapes. After the piston reaches the bottom of the barrel, it begins its ascent. This would cause a vacuum between the bottom of the barrel and the ascending piston (since the unbalanced pressure of the outside air immediately closes the valve *s*), but the tension of the air in the receiver R opens the valve *t* and fills this space. As the air in R expands it becomes rarefied and loses some of its tension. The external pressure of the air on R, being no longer balanced by the tension of the air within, presses the receiver firmly upon the plate L. Each repetition of a double stroke of the piston removes a portion of the air remaining in R. The air is removed from R by its own expansion. However far the process of exhaustion may be carried, the receiver will always be filled with air, although it may be exceedingly rarefied. The operation of exhaustion is practically ended when the tension of the air in R becomes too feeble to lift the valve *t*.

Sometimes another receiver, D, is used, opening into the tube T, that connects the receiver with the barrel. Inside the receiver is placed a barometer. It is apparent that air is exhausted from D as well as from R; and, as the pressure is removed from the surface of the mercury in the cup, the bar-

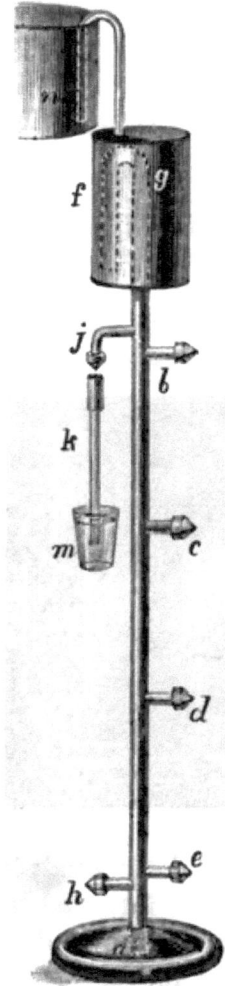

Fig. 34.

ometric column falls; so that the barometer serves as a gauge to indicate the approximation to a vacuum. For instance, when the mercury has fallen 380^{mm} (15 inches), one-half of the air has been removed.

41. Sprengel Pump.

Experiment 38. — Remove the cap from j (Fig. 34), and connect with a glass tube k, about 12 inches long. Let k dip into a tumbler of water, m. Support the apparatus on a couple of blocks of wood, so that when the stopper a in the base is removed, the water may fall freely out at the bottom. Fill the cup g with water, and allow it to escape at a. As the water passes the branch tube j, the expansive air in the tube gets entangled in the water, and is constantly removed by the falling stream, and thus a partial vacuum is formed in the tube k. The pressure of air on the surface of the water in the open cup forces the water up the tube k, and empties the tumbler. If m were a closed vessel filled with air, it is apparent that a partial vacuum would be created in it. An apparatus constructed like this, in which mercury is employed instead of water, constitutes one of the most efficient air-pumps in use. It is called the Sprengel pump.

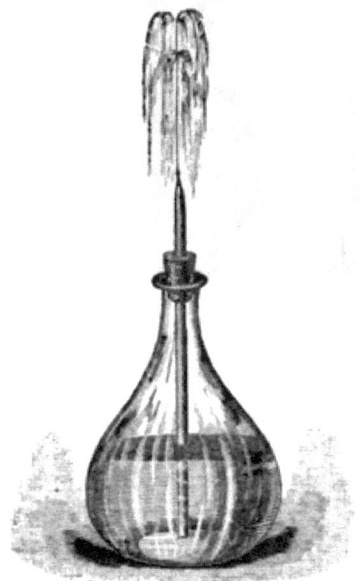

Fig. 35.

Modifications of this pump have extensive use in the arts, such as in obtaining high vacua in electrical lamps, radiometers, etc. By means of a good Sprengel pump exhaustion to the hundred-millionth of an atmosphere can be attained. In such a space it is calculated that a molecule of air traverses an average distance of 33 feet before colliding with another molecule of air.

44 DYNAMICS OF FLUIDS.

42. Condenser.

Experiment 39.—Into the neck of a bottle partly filled with water (Fig. 35) insert a cork very tightly, through which pass a glass tube nearly to the bottom of the bottle. Blow forcibly into the bottle. On removing the mouth water will flow through the tube in a stream. Explain.

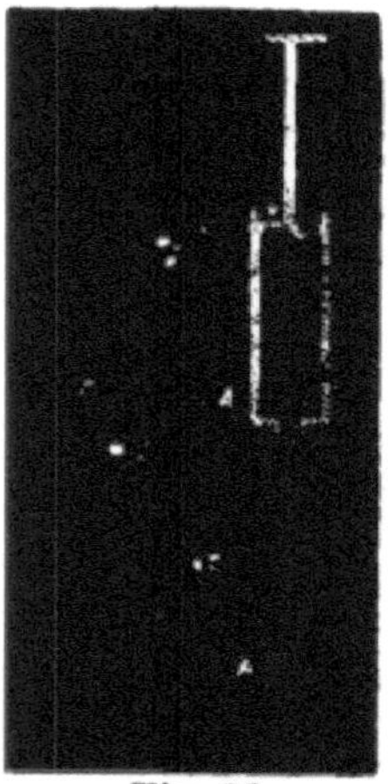

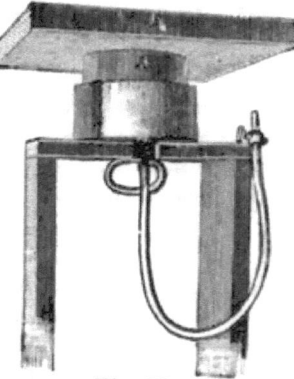

Figure 6, page 5, represents in perspective, and Figure 36, in section, an apparatus for condensing air, called a *condenser*. Its construction is like that of the barrel of an air-pump, except that the direction in which the valves open is reversed.

Fig. 36.
Fig. 37.

Experiment 40.—Place a block having a wide platform at one end on the piston of the seven-in-one apparatus. On the platform let a child stand. By means of a condensing syringe (Fig. 6), connected by a rubber tube with the seven-in-one apparatus (Fig. 37), condense the air in the cylinder and raise the child.

Section V.

APPARATUS FOR RAISING LIQUIDS.

43. Lifting or Suction Pump.—The common *lifting-pump* is constructed like the barrel of an air-pump. Figure 38 represents the piston B in the act of rising. As the air is rarefied below it, water rises in consequence of atmospheric pressure on the water in the well, and opens the lower valve D. Atmospheric pressure closes

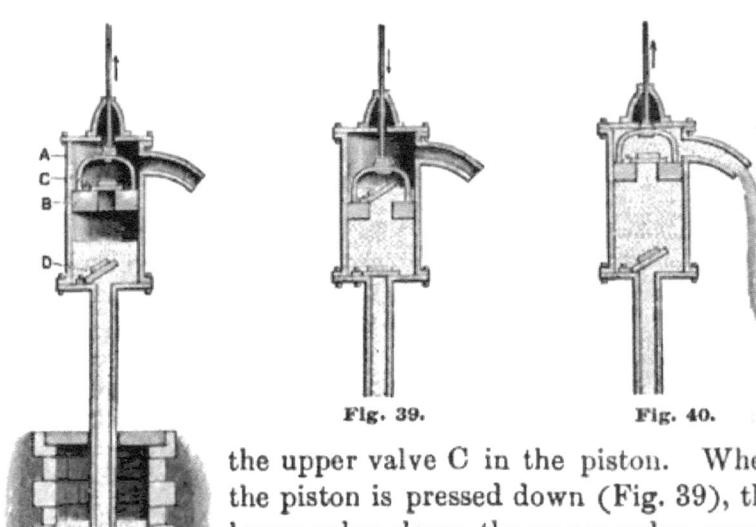

Fig. 39.   Fig. 40.

Fig. 38.

the upper valve C in the piston. When the piston is pressed down (Fig. 39), the lower valve closes, the upper valve opens, and the water between the bottom of the barrel and the piston passes through the upper valve above the piston. When the piston is raised again (Fig. 40), the water above the piston is raised and discharged from the spout.

The liquid is sometimes said to be raised in a lifting-pump by the "force of suction." Is there such a *force?*

Fig. 41.

Experiment 41. — Bend a glass tube into a U-shape, with unequal arms, as in Figure 41. Fill the tube with the liquid to the level cb. Close the end b with a finger, and try to suck the liquid out of the tube. You find it impossible. Remove the finger from b, and you can suck the liquid out with ease. Why?

DYNAMICS OF FLUIDS.

44. Force-Pump. — The piston of a *force-pump* (Fig. 42) has no valve, but a branch pipe *a* leads from the lower part of the barrel to an air-condensing chamber *b*, at the bottom of which is a valve *c*, opening upward. As the piston is raised, water is forced up through the valve *d*, while water in *b* is prevented from returning by the valve *c*. When the piston is forced down, the valve *d* closes, the valve *c* opens, and the water is forced into the chamber *b*, condensing the air above the water. The elasticity of the condensed air forces the water out of the tube *e* in a continuous stream.

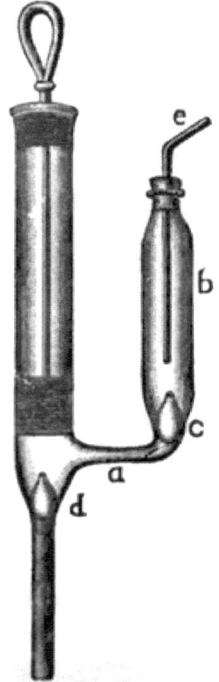

Fig. 42.

QUESTIONS AND PROBLEMS.

1. What force is the cause of fluid pressure?
2. Why does not a person at the bottom of a pond feel the weight of the water above him?
3. An aeronaut finds that on the earth his barometer stands at 30 inches. He ascends in a balloon until the barometer stands at 20 inches. About how high is he? What is the pressure of the atmosphere at his elevation?
4. When a barometer stands at 30 inches, the atmospheric pressure is 14.7 pounds. What is the atmospheric pressure when the barometer stands at 29 inches?
5. Why is a barometer tube closed at the top? Why must air come in contact with the mercury at the bottom?
6. What would be the effect on an aneroid barometer if it were placed under the receiver of an air-pump, and one or two strokes of the pump were made?
7. Suppose a rubber foot-ball to be partially inflated with air at the surface of the earth; what would happen if it were taken up in a balloon?
8. Mercury is 13.6 times denser than water. When a mercurial ba-

rometer stands at 30 inches, how high would a water barometer stand? How high, theoretically, could mercury be raised on such a day by suction? How high could water be raised by the same means? How many times higher can water be raised by a suction-pump than mercury?

9. What is that which is sometimes called the "force of suction"?

10. The area of one side of the piston of the seven-in-one apparatus is about 26 square inches. Suppose the piston to be forced into the cylinder so as to drive out all the air, and then the orifice to be closed; what force would be required to draw the piston out, when the barometer stands at 30 inches? What force would be required on the top of a mountain where the barometer stands at 15 inches?

11. Water is raised the larger part of the distance in our lifting-pumps by atmospheric pressure; why, then, is not such a pump a labor-saving instrument?

12. If water is to be raised from a well 50 feet deep, how high must it be lifted, and how long must the barrel be?

Section VI.

TRANSMISSION OF EXTERNAL PRESSURE.

45. Pressure Transmitted Undiminished in All Directions.

Experiment 42. — Fill the glass globe and cylinder (Fig. 43) with water, and thrust the piston into the cylinder. Jets of water will be thrown not only from that aperture a in the globe toward which the piston moves and the pressure is exerted, but from apertures on all sides. Furthermore, the streams extend to equal distances in every direction.

It thus appears that external pressure is exerted not alone upon that portion of the liquid that lies in the path of the force, but it is transmitted equally to all parts and in all directions.

48 DYNAMICS OF FLUIDS.

Experiment 43.— Measure the diameter of the bore of each arm of the glass U-tube (Fig. 44). We will suppose, for illustration, that the diameters are respectively 40^{mm} and 10^{mm}; then the areas of the transverse sections of the bores will be $40^2 : 10^2 = 16$; that is, when the tube contains a liquid, the area of the free surface of the liquid in the large arm will be 16 times as great as that in the small arm. Pour mercury into the tube until it stands about 1^{cm} above the bottom of the large arm. The mercury stands at the same level in both arms. Pour water upon the mercury in the large arm until this arm lacks only about 1^{cm} of being full. The pressure of the water causes the mercury to rise in the small arm, and to be depressed in the large arm. Pour water very slowly into the small arm from a beaker having a narrow lip, until the surfaces of the water in the two arms are on the same level. It is evident that the quantity of water in the large arm is 16 times as great as that in the small arm. This phenomenon appears paradoxical (*apparently* contrary to the natural course of things), until we master the important hydrostatic principle involved. We must not regard the body of mercury as serving as a balance beam between the two bodies of water, for this would lead to the absurd conclusion that a given mass of matter may balance another mass 16 times as great. We may best understand this phenomenon by imagining the body of liquid in the large arm to be divided into cylindrical columns of liquid of the same size as that in the small arm. There will evidently be 16 such columns. Then whatever pressure is exerted on the mercury by the water in the small arm is transmitted by the mercury to each of the 16 columns, so that each column receives an upward pressure, or a supporting force equal to the weight of the water in the small arm. This method of transmit-

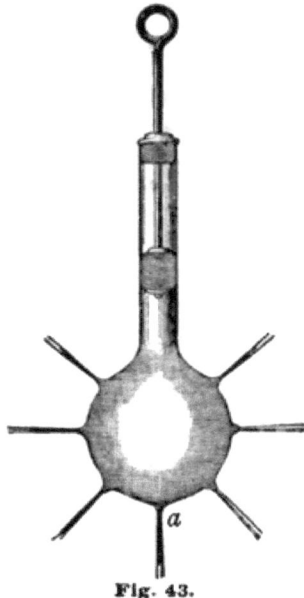

Fig. 43.

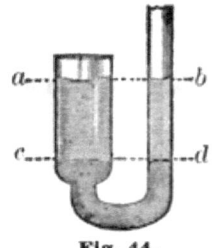

Fig. 44.

ting pressure is peculiar to fluids. With solids it is quite different. If the mercury in our experiment were a solid body, it would require equal masses of water placed upon the two extremities to counterbalance each other.

Experiment 44.— Support the seven-in-one apparatus with the open end upward, force the piston in, and place on it a block of wood A (Fig. 45), and on the block a heavy weight (or let a small child stand on the block). Attach one end of the rubber tube B (12 feet long) to the apparatus, and insert a tunnel C in the other end of the tube. Raise the latter end as high as practicable, and pour water into the tube. Explain how the few ounces of water standing in the tube can exert a pressure of many pounds on the piston, and cause it to rise together with the burden that is on it.

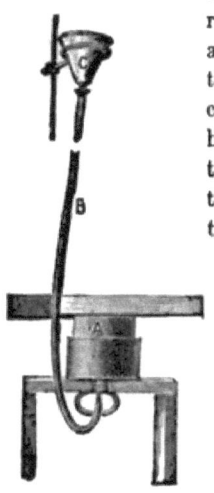

Fig. 45.

Fig. 46.

Experiment 45.— Remove the water from the apparatus, place on the piston a 16-pound weight, and blow (Fig. 46) from the lungs into the apparatus. Notwithstanding that the actual pushing force exerted through the tube by the lungs does not probably exceed an ounce, the slight increase of tension caused thereby when exerted upon the (about) 26 square inches of surface of the piston causes it to rise together with its burden.

A pressure exerted on a given area of a fluid enclosed in a vessel is transmitted to every equal area of the interior of the vessel; and the whole pressure that may be exerted upon the vessel may be increased in proportion as the area of the part subjected to external pressure is decreased.

46. Hydrostatic Press. — This principle has an important practical application in the *hydrostatic press.* You see two pistons t and s (Fig. 47). The area of the lower surface of t is (say) one hundred times that of the lower surface of s.

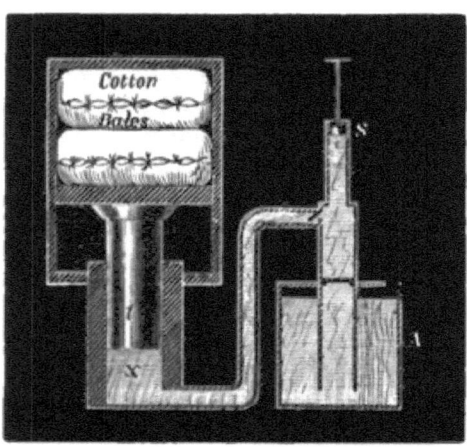

Fig. 47.

As the piston s is raised and depressed, water is pumped up from the cistern A, forced into the cylinder x, and exerts a total upward pressure against the piston t one hundred times greater than the downward pressure exerted upon s. Thus, if a pressure of one hundred pounds is applied at s, the cotton bales will be subjected to a pressure of five tons.

The pressure that may be exerted by these presses is enormous. The hand of a child can break a strong iron bar. But observe that, although the pressure exerted is very great, the upward movement of the piston t is very slow. In order that the piston t may rise 1 inch, the piston s must descend 100 inches. The disadvantage arising from slowness of operation is little thought of, however, when we consider the great advantage accruing from the fact that one man can produce as great a pressure with the press as a hundred men can exert without it.

The press is used for compressing cotton, hay, etc., into bales, and for extracting oil from seeds. The modern engineer finds it a most efficient machine, whenever great weights are to be moved through short distances, as in launching ships.

Section VII.

PRESSURE EXERTED BY LIQUIDS DUE TO THEIR OWN WEIGHT.

47. Pressure Dependent on Depth, but Independent of the Quantity and Shape of a Body of Liquid. — Having considered the transmission of *external pressure* applied to any portion of a liquid, we proceed to examine the effects of pressure due to the weight of liquids themselves.

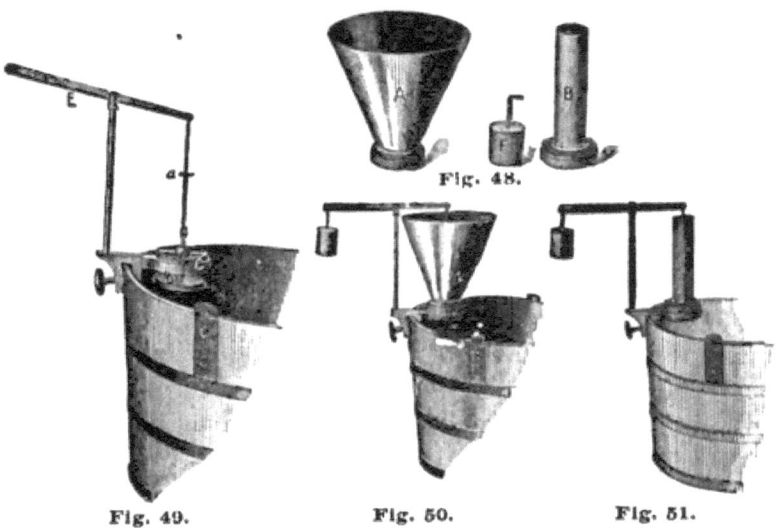

Fig. 48.

Fig. 49.   Fig. 50.   Fig. 51.

Experiment 46. — A and B (Fig. 48) are two bottomless vessels which can be alternately screwed to a supporting ring C (Fig. 49). The ring is itself fastened by means of a clamp to the rim of a wooden water-pail. A circular disk of metal, D, is supported by a rod connected with one arm of the balance-beam E. When the weight F is applied to the other arm of the beam, the disk D is drawn up against the ring so as to supply a bottom for the vessel above. Take first the vessel A, screw it to the ring, and apply the weight to the beam as in Figure 50. Pour water slowly into the vessel, moving the index *a* up the rod so as to keep it just at the surface of the water, until the downward pressure of the water upon the bottom tilts the beam, and pushes the bottom down from the ring, and allows some of the water to fall into the pail. Remove vessel A, and attach B to the ring as in Figure 51. Pour water as before into vessel B; when the surface of the water reaches the index a, the bottom is forced off as before. That is, *at the same depth, though the quantity of water and the shape of the vessel be different, the pressure upon the bottom of a vessel is the same, provided the bottom is of the same area.*

48. Rules for Calculating Liquid Pressure against the Bottom and Sides of a Containing Vessel. — *The pressure due to gravity on any portion of the bottom of a vessel containing a liquid is equal to the weight of a column of the same liquid whose base is the area of that portion of the bottom pressed upon, and whose hight is the greatest depth of the water in the vessel.* Thus, suppose that we have three vessels having bottoms of the same size: one of them has flaring sides, like a wash-basin; another has cylindrical sides; and the third has conical sides, like a coffee-pot. If the three vessels are filled with water to the same depth, the pressure upon the bottom of each will be equal to the weight of the water in the vessel of cylindrical shape. Suppose that the area of the bottom of each is 108 square inches, and the depth of water is 16 inches; then the cubical contents of the water in the cylindrical vessel is 1,728 cubic inches, or 1 cubic foot. The weight of 1 cubic foot of water is $62\frac{1}{2}$ pounds. Hence, the pressure upon the bottom of each vessel is $62\frac{1}{2}$ pounds.

Evidently, the lateral pressure at any point of the side of a vessel depends upon the depth of that point; and, as depth at different points of a side varies, hence, *to find the pressure upon any portion of a side of a vessel, we find the weight of a column of liquid whose base is the area of that portion of the side, and whose hight is the average depth of that portion.*

49. The Surface of a Liquid at Rest is Level. — This fact is commonly expressed thus: "Water always seeks its lowest level." In accordance with this principle, water flows down an inclined plane, and will not remain heaped up. An illustration of the application of this principle, on a large scale, is found in the method of supplying cities with water. Figure 52 represents a modern aqueduct, through which water is conveyed from an elevated pond or river *a*, beneath a river *b*, over a hill *c*, through a valley

Fig. 52.

d, to a reservoir *e*, in a city, from which water is distributed by service-pipes to the dwellings. The pipe is tapped at different points, and fountains at these points would rise to the level of the water in the pond, but for the resistance of the air, friction in the pipes, and the check which the ascending steam receives from the falling drops. Where should the pipes be made stronger, on a hill or in a valley? Where will water issue from faucets with greater force, in a chamber or in a basement? How high may water be drawn from the pipe in the house *f*?

Section VIII.

THE SIPHON.

50. Construction and Operation of the Siphon. — A siphon is an instrument used for transferring a liquid from one vessel to another through the agency of atmospheric pressure. It consists of a tube of any material (rubber is often most convenient) bent into a shape somewhat like the letter U. To set it in operation, fill the tube with a liquid, stop each end with a finger or cork, place it in the position represented in Figure 53, remove the stoppers and the liquid will all flow out at the orifice o. Why? The upward pressure of the atmosphere against the liquid in the tube is the same at both ends; hence these two forces are in equilibrium. But the weight of the column of liquid ab is greater than the weight of the column dc; hence equilibrium is destroyed and the movement is in the direction of the greater (*i.e.* the unbalanced) force. The unbalanced force which causes the flow is equal to the weight of the column eb.

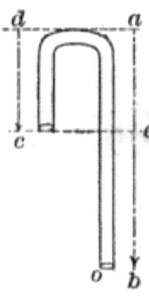

Fig. 53.

If one end of the tube filled with liquid is immersed in a liquid in some vessel, as in A, Figure 54, and the other end is brought below the surface of the liquid in the vessel and the stoppers are removed, the liquid in the vessel will flow out through the tube until the distance eb becomes zero.

If one of the vessels is raised a little, as in C, the liquid will flow from the raised vessel, till the surfaces in the two vessels are on the same level.

THE SIPHON. 55

The remaining diagrams in this cut represent some of the great variety of uses to which the siphon may be put. D, E, and F are different forms of siphon fountains. In D, the siphon tube is filled by blowing in the tube *f*. Explain the remainder of the operation. A siphon of the form G is always ready for use. It is only necessary to dip one end into the liquid to be

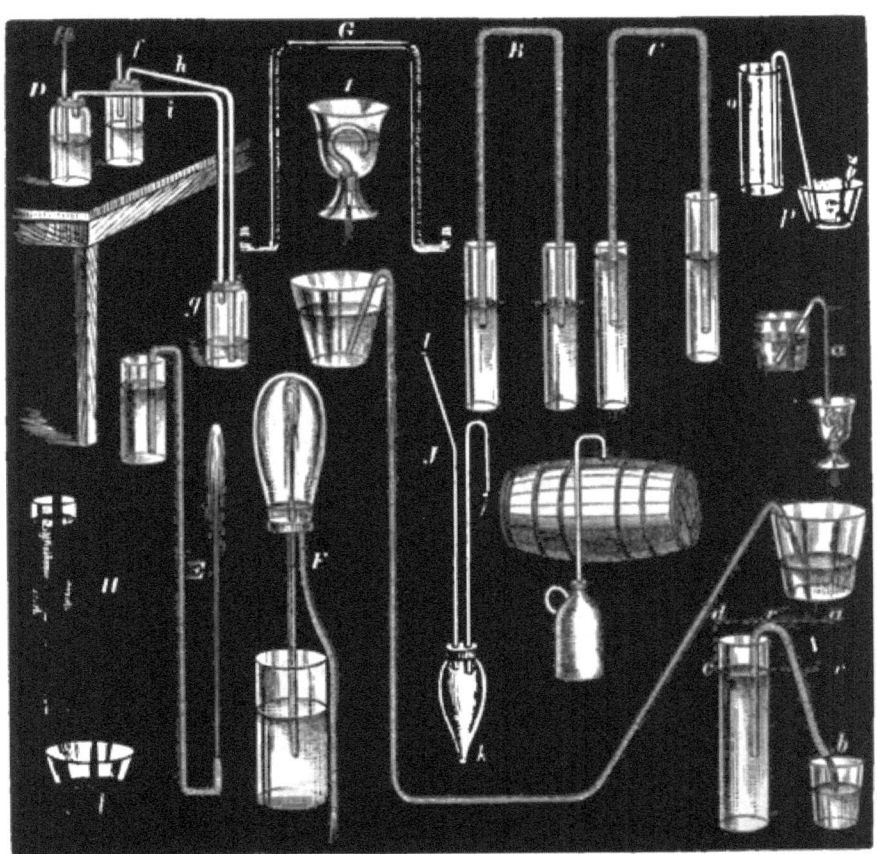

Fig. 54.

transferred. Why does the liquid not flow out of this tube in its present condition? H illustrates the method by which a heavy liquid may be removed from beneath a lighter liquid. By means of a siphon a liquid may be removed from a vessel in a clear state, without disturbing sediment at the bottom. I is a *Tantalus Cup*. A liquid will not flow from this cup till the top of the bend of the tube is covered. It will then continue to flow as long as the end of the tube is in the liquid. The cup g (Fig. 34, page 42) is a Tantalus cup. The siphon J may be filled with a liquid that is not safe or pleasant to handle, by placing the end j in the liquid, stopping the end k, and sucking the air out at the end l till the lower end is filled with the liquid.

Gases heavier than air may be siphoned like liquids. Vessel o contains carbonic-acid gas. As the gas is siphoned into the vessel p, it extinguishes a candle-flame. Gases lighter than air are siphoned by inverting both the vessels and the siphon.

Section IX.

BUOYANT FORCE OF FLUIDS.

51. Origin of Buoyancy.

Experiment 47. — Gradually lower a large stone, by a string tied to it, into a bucket of water, and notice that its weight gradually becomes less till it is completely submerged. Slowly raise it out of the water, and note the change in weight as it emerges from the water. Suspend the stone from a spring balance, weigh it in air and then in water, and ascertain its loss of weight in the latter.

Fig. 55.

It seems as if something in the fluid, underneath the articles submerged, were pressing up against them. A moment's reflection will make the explanation of this phenomenon apparent. We have learned (1) that pressure at any given point in a body of fluid is equal in all directions. (2) That pressure in liquids increases as the depth. Consequently, the downward pressure on the *top* (*i.e.* the place of least depth) of a body immersed in a fluid, as *dcba* (Fig. 55), must be less than the upward pressure against the bottom; hence, there is an unbalanced force acting upward, which tends to neutralize to some extent the weight or gravity of the body. This unbalanced force is called the *buoyant force* of fluids. That there is equilibrium between the pressures on the sides of a body immersed is shown by the fact that there is no tendency to move laterally.

52. Magnitude of the Buoyant Force.

Experiment 48. — Suspend from one arm of a balance beam a cylindrical bucket A (Fig. 56), and from the bucket a solid cylinder whose volume is exactly equal to the capacity of the bucket; in other words, the latter would just fill the former. Counterpoise the bucket and cylinder with weights.

Place beneath the cylinder a tumbler of water, and raise the tumbler until the cylinder is completely submerged. The buoyant force of the water destroys the equilibrium. Pour water into the bucket; when it becomes just even full, the equilibrium is restored.

Now it is evident that the cylinder immersed in the water displaces its own volume of water, or just as much water as fills the bucket. But the bucket full of water is just sufficient to restore the weight lost by the submersion of the cylinder. Hence, *a solid immersed in a liquid is buoyed up with a force equal to* (*i.e.* its apparent loss in weight is) *the weight of the liquid it displaces.*

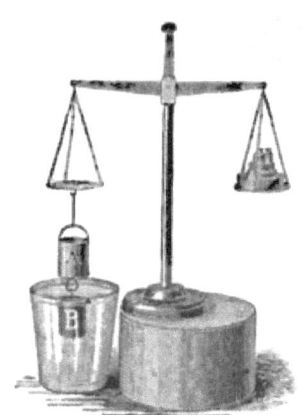

Fig. 56.

Experiment 49. — The last statement may be verified in another way with apparatus like that shown in Figure 57. Fill the vessel A till the liquid overflows at E. After the overflow ceases, place a vessel c under the nozzle. Suspend a stone from the balance-beam B, and weigh it in air, and then carefully lower it into the liquid,

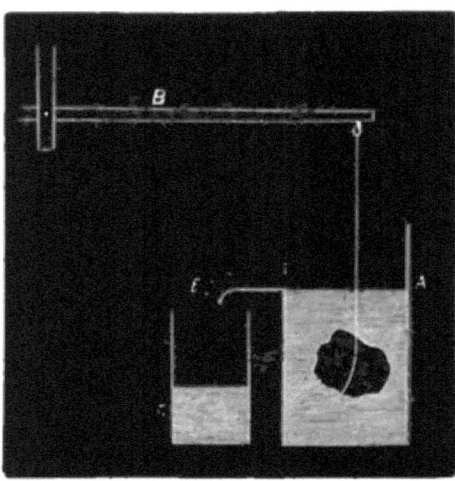

Fig. 57.

when some of the liquid will flow into the vessel c. The vessel c having been weighed when empty, weigh it again with its liquid contents, and it will be found that its increase in weight is just equal to the loss of weight of the stone.

Experiment 50. — Next suspend a block of wood that will float in the liquid, and weigh it in air. Then float it upon the liquid, and weigh the liquid displaced as before, and it will be found that the weight of the liquid displaced is just equal to the weight of the block in air.

Hence, *a floating body displaces its own weight of liquid; in other words, a floating body will sink till it displaces an equal weight of the liquid, or till it reaches a depth where the buoyant force is equal to its own weight.*

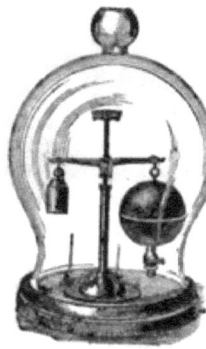

Fig. 58.

Experiment 51. — Place a baroscope (Fig. 58), consisting of a scale-beam, a small weight, and a hollow brass sphere, under the receiver of an air-pump, and exhaust the air. In the air the weight and sphere balance each other; but when the air is removed, the sphere sinks, showing that in reality it is heavier than the weight. In the air each is buoyed up by the weight of the air it displaces; but as the sphere displaces more air, it is buoyed up more. Consequently, when the buoyant force is withdrawn from both, their equilibrium is destroyed.

We see from this experiment that *bodies weigh less in air than in a vacuum*, and that *we never ascertain the true weight of a body, except when weighed in a vacuum.*

The density of the atmosphere is greatest at the surface of the earth. A body free to move cannot displace more than its own weight of a fluid; therefore a balloon, which is a large bag filled with a gas about fourteen times lighter than air at the sea-level, will rise till the balloon, plus the weight of the car and cargo, equals the weight of the air displaced.

Figure 59 represents a water-tank in common use in our houses. Water enters it from the main until nearly full, when it reaches the hollow metallic ball A, and raises it by its buoyant force and closes a valve in the main pipe, and thus prevents an overflow. An overflow is still further prevented by the waste pipe and another "ball tap," B, which opens at a suitable time another passage for the escape of water.

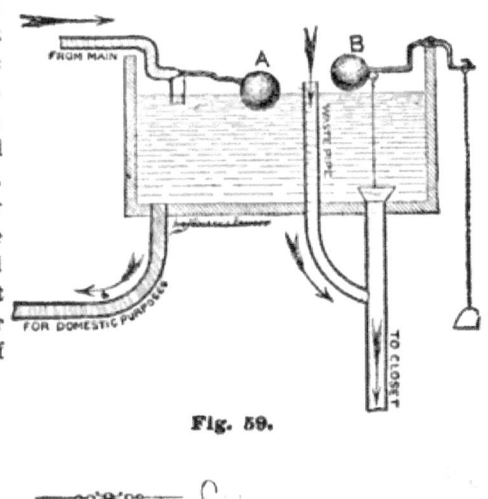

Fig. 59.

Section X.

DENSITY AND SPECIFIC GRAVITY.

53. Meaning of the Terms and their Relation to each Other. — The quantity of matter per unit of volume represents the *density* of the matter filling that space.

Thus, a gram of water at 4° C. (centigrade thermometer) occupies a cubic centimeter; while the same space would contain 11.5 grams of lead. Every kind of matter (*i.e.* every substance) has a special or *specific* density of its own. Pure water at 4° C. is taken as a standard; and its density is said to be $\left(\dfrac{\text{mass}}{\text{volume}} = \dfrac{1^g}{1^{cc}} = \right) 1$. In the same way the density of lead is $\left(\dfrac{11.5^g}{1^{cc}} = \right) 11.5$. A piece of lead which occupies a given space not only contains 11.5 times as much matter, but also *weighs* 11.5 times as much as the quantity of water which would fill the same space. The density of any liquid or solid compared with that of water is a ratio — called its *specific density*; this ratio is numerically equal to the ratio, called its *specific gravity*, of its weight compared with the weight of an equal volume of water at the standard temperature.

54. Formulas for Specific Density and Specific Gravity. — Let D represent the density of any given substance (*e.g.* lead), and D the density of water, and let G and G′ represent respectively the weights of equal volumes of the same substances; then

(1) $\dfrac{\text{Density of given substance}}{\text{Density of water}} = \dfrac{D}{D'} = \text{Sp. D.}$

(2) $\dfrac{\text{Weight of a given volume of the substance}}{\text{Weight of equal volume of water}} = \dfrac{G}{G'} = \text{Sp. G.}$

The Sp. D. of lead $= \dfrac{D}{D'} = \dfrac{11.5}{1} = 11.5$. The Sp. G. of lead $= \dfrac{G}{G'} = \dfrac{11.5}{1} = 11.5$. Hence Sp. D. and Sp. G. are numerically equal. In the same way ratios may be found for other substances and recorded in a table; such a table exhibits both the specific densities and the specific gravities of the substances. See Appendix B.

Section XI.

EXPERIMENTAL METHODS OF FINDING THE SPECIFIC DENSITY AND SPECIFIC GRAVITY OF BODIES.

55. Solids.

Experiment 52. — From a hook beneath a scale-pan (Fig. 60) suspend by a fine thread a small specimen of a substance whose specific gravity is to be found, and weigh it, while dry, in the air. Then immerse the body in a tumbler of water (do not allow it to touch the tumbler, and see that it is completely submerged), and weigh it in water. The loss of weight in water is evidently G', *i.e.* the weight of the water displaced by the body; or, in other words, the weight of a body of water having the same volume as that of the specimen. Apply the formula (2) for finding the specific gravity.

Fig. 60.

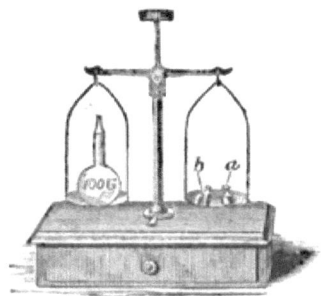

Fig. 61.

Experiment 53. — Take a piece of sheet lead one inch long and one-half inch wide, weigh it in air and then in water, and find its loss of weight in water. [It will not be necessary to repeat this part of the operation in future experiments.] Weigh in air a piece of cork or other substance that floats in water, then fold the lead-sinker, and place it astride the string just above the specimen, completely immerse both, and find their combined weight in water. Subtract their combined weight in water from the sum of the weights of both in air; this gives the weight of water displaced by both. Subtract from this the weight lost by the lead alone, and the remainder is G'; *i.e.* the weight of water displaced by the cork. Apply formula (2), as before.

56. Liquids.

Experiment 54. — Take a specific-gravity bottle that holds when filled a certain (round) number of grams of water, *e.g.* 100^g, 200^g, etc. Fill the bottle with the liquid whose specific gravity is sought. Place it on a scale-pan (Fig. 61), and on the other scale-pan place a piece of metal *a*, which is an exact counterpoise for the bottle when empty. On the same pan place weights *b*, until there is equilibrium. The weights placed in this pan represent the weight W of the liquid in the bottle. Apply formula (2). The W' (*i.e.* the 100^g, 200^g, etc.) is the same in every experiment, and is usually etched on the bottle.

Experiment 55. — Take a pebble stone (*e.g.* quartz) about the size of a large chestnut, find its loss of weight (*i.e.* W') in water; find its loss of weight (*i.e.* W) in the given liquid. Apply formula (2).

Prepare blanks, and tabulate the results of the experiments above as follows: —

Name of Substance.	W in Grams.	W' in Grams.	Sp. G. or Sp. D.	E.
Lead	7.2	6.6	12	.5

When the result obtained differs from that given in the table of specific gravities (see Appendix B), the difference is recorded in the column of errors (*e*). The results recorded in the column of errors are not necessarily *real* errors; they may indicate the degree of impurity, or some peculiar physical condition, of the specimen tested.

57. Hydrometers.

— If a wooden, an iron, and a lead ball are placed in a vessel containing mercury (Fig. 62),

SPECIFIC GRAVITY AND SPECIFIC DENSITY. 63

they will float on the mercury at different depths, according to their relative densities. Ice floats, in water with $\frac{918}{1000}$, in mercury with $\frac{60}{1000}$, of its bulk submerged. Hence the Sp. D. of mercury is $918 \div 60 =$ about 13.5.

We see, then, that the densities of liquids may be compared by seeing to what depths bodies floating in them will sink. An instrument (A, Fig. 63) called a *hydrometer* is constructed on this principle. It consists of a glass tube with one or more bulbs blown in it, loaded at one end with shot or mercury to keep it in a vertical position when placed in a liquid. It has a scale of specific densities on the stem, so that the experimenter has only to place it in the liquid to be tested, and read its specific density or specific gravity at that point, B, of the stem which is at the surface of the liquid.

Fig. 62.

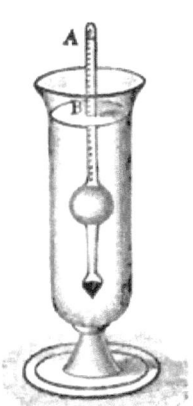

Fig. 63.

58. Miscellaneous Experiments.

Experiment 56.—Find the cubical contents of an irregular shaped body, *e.g.* a stone. Find its loss of weight in water. Remember that the loss of weight is precisely the weight of the water it displaces, and that the volume of one gram of water is one cubic centimeter.

Experiment 57. — Find the capacity of a test-tube, or an irregular shaped cavity in any body. Weigh the body; then fill the cavity with water, and weigh again. As many grams as its weight is increased, so many cubic centimeters is the capacity of the cavity.

Experiment 58. — A fresh egg sinks in water. See if by dissolving table salt in the water it can be made to float. How does salt affect the density of the water?

Experiment 59. — Float a sensitive hydrometer in water at about 60° F. (15° C.), and in other water at about 180° F. (82° C.). Which water is denser?

EXERCISES.

1. In which does a liquid stand higher, in the snout of a coffee-pot or in the main body? On which does this show that pressure depends, on quantity or depth of liquid?

2. The areas of the bottoms of vessels A, B, and C (Fig. 64) are equal. The vessels have the same depth, and are filled with water. Which vessel contains the more water? On the bottom of which vessel is the pressure equal to the weight of the water which it contains? How does the pressure upon the bottom of vessel B compare with the weight of the water in it?

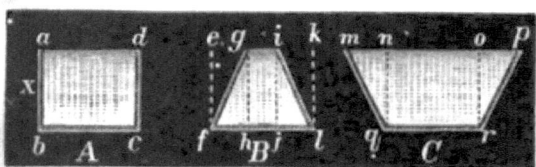

Fig. 64.

3. A cubic foot of water weighs about 62.5 pounds or 1,000 ounces. Suppose that the area of the bottom of each vessel is 50 square inches and the depth is 10 inches; what is the pressure on the bottom of each?

4. Suppose that the vessel A is a cubical vessel; what is the pressure against one of its vertical sides?

5. Suppose that vessel A were tightly covered, and that a tube 10 feet long were passed through a perforation in the cover so that the end just touches the upper surface of the water in the vessel; then suppose the tube to be filled with water. If the area of the cross-section

SPECIFIC GRAVITY AND SPECIFIC DENSITY.   65

of the bore is 1 square inch, what additional pressure will each side of the cube sustain?

6. Suppose that the area of the end of the large piston of a hydrostatic press is 100 square inches; what should be the area of the end of the small piston that a force of 100 pounds applied to it may produce a pressure of 2 tons?

7. A solid body weighs 10 pounds in air and 6 pounds in water. (*a*) What is the weight of an equal bulk of water? (*b*) What is its specific gravity? (*c*) What is the volume of the body? (*d*) What would it weigh if it were immersed in sulphuric acid? [See table of specific gravities, Appendix B.]

8. A thousand-grain specific-gravity bottle filled with sea-water requires in addition to the counterpoise of the bottle 1,026 grains to balance it. (*a*) What is the specific gravity of sea-water? (*b*) What is the quantity of salt, etc., dissolved in 1,000 grams of sea-water?

9. A piece of cork floating on water displaces 2 pounds of water. What is the weight of the cork?

10. In which would a hydrometer sink farther, in milk or water?

11. What metals will float in mercury?

12. (*a*) Which has the greater specific gravity, water at 10° C. or water at 20° C.? (*b*) If water at the bottom of a vessel could be raised by application of heat to 20° C. while the water near the upper surface has a temperature of 10° C., what would happen?

13. A block of wood weighs 550 grams; when a certain irregular-shaped cavity is filled with mercury the block weighs 570 grams. What is the capacity or cubical contents of the cavity?

14. In which is it easier for a person to float, in fresh water or in sea-water? Why?

15. Figure 65 represents a beaker graduated in cubic centimeters. Suppose that when water stands in the graduate at 50^{cc}, a pebble stone is dropped into the water, and the water rises to 75^{cc}. (*a*) What is the volume of the stone? (*b*) How much less does the stone weigh in water than in air? (*c*) What is the weight of an equal volume of water?

16. If a piece of cork is floated on water in a graduate, and displaces (*i.e.* causes the water to rise) 7^{cc}, what is the weight of the cork?

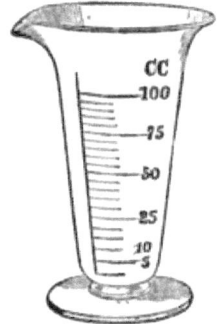

Fig. 65.

17. If a piece of lead (sp. g. 11.35) is dropped into a graduate and displaces 12^{cc} of water, what does the lead weigh? (*a*) How would you measure out 50 grams of water in a graduate? (*b*) How would you measure out the same weight of alcohol (sp. g. 0.8)? (*c*) How the same weight of sulphuric acid (sp. g. 1.84)?

18. What is the density of gold? silver? milk? alcohol?

19. When the barometer stands at 30 inches, how high can alcohol be raised by a perfect lifting-pump?

20. A measuring glass graduated in cubic centimeters contains water. An empty bottle floats on the water, and the surface of the water stands at 50^{cc}. If 10^g of lead shot are placed in the bottle, where will the surface of the water stand?

21. What evidence do we see daily that there is relative motion between the sun and the earth?

22. On what two things does the weight of a body depend?

23. (*a*) Can you suck air out of a bottle? (*b*) Can you suck water out of a bottle? Explain.

24. (*a*) What bodies have neither volume nor shape? (*b*) What have volume, but not shape? (*c*) What have both volume and shape?

25. When the volume of a body of gas diminishes, is it due to contraction or compression, *i.e.* to internal or external forces?

26. What is the hight of the barometer column when the atmospheric pressure is 10 grams per square centimeter?

27. A barometer in a diving-bell (page 3) stands at 96^{cm} when a barometer at the surface of the earth stands at 76^{cm}; what is the depth of the surface of water inside the bell?

CHAPTER III.

GENERAL DYNAMICS.

Section I.

MOMENTUM AND ITS RELATION TO FORCE.

59. Momentum. — An empty car in motion is much more easily stopped than a loaded car moving with the same speed. Evidently, if force is employed to destroy motion, and it takes either a greater force to stop the loaded car in a given time, or the same force a longer time, it follows that there must be *more motion to be destroyed* in the loaded car than in the empty car moving with the same velocity. Quantity of motion, more briefly *momentum*, and velocity are not identical. Momentum depends upon both *mass* and *velocity*; velocity is independent of mass. Momentum $= MV$.

The momentum of a moving body is measured by the product of its mass multiplied by its velocity.

60. Relation of Momentum to Force.

Experiment 60. — Weights A and B of the Atwood machine (Fig. 66), suspended by a thread passing over the wheel C, are in equilibrium with reference to the force of gravity; consequently neither falls. Raise weight A, and let it rest on the platform D, as in Figure 67. The two weights are still in equilibrium. Place weight E, called a "rider," on A. There is now an unbalanced force, and if the platform D is removed, there will be motion, *i.e.* A and E will fall, and B will rise. Set the pendulum F to vibrating. At each vibration it

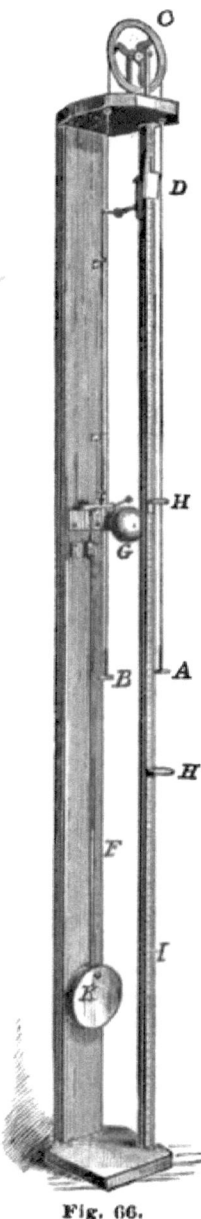

Fig. 66.

causes a stroke of the hammer on the bell G. At the instant of the first stroke the pendulum causes the platform D to drop so as to allow the weights to move. When the weights reach the ring H, the rider is caught off by the ring.

Raise and lower the ring on the graduated pillar I, and ascertain by repeated trials the average distance the weights descend in the interval between the first two strokes of the bell.

Next substitute for E a weight L, double that of E. Find by trial how far the weights now descend in the same interval of time as before. It will be found that in the latter case the weights descend nearly twice as far as in the first case.

Suppose that weights A and B are each 30 grams, and that weights E and L are respectively 2 grams and 4 grams. Now the force of gravity which acts on weight E is 2 grams. Consequently the unbalanced force which put in motion the three weights A, B, and E, whose combined weight (disregarding the weight of wheel C, which is also put in motion) is $(30 + 30 + 2 =)$ 62 grams, was 2 grams. It is now evident why the descent is slow, for instead of a force of 1 gram acting upon each gram of matter, as is usually the case with falling bodies, we have a force of only 2 grams moving 62 grams of matter; consequently the descent is about $\frac{1}{31}$ as fast as that of falling bodies generally.

But when we employed weight L, we had a force of 4 grams moving $(30 + 30 + 4 =)$ 64 grams of matter. Here the force is doubled, and the distance traversed is nearly doubled; consequently the average velocity and the momentum acquired are *nearly* doubled. Had the masses moved in the two cases been exactly the same, the velocity and the momentum would have been exactly doubled.

(1) *In equal intervals of time change of momentum is proportional to the force employed.*

Experiment 61. — Once more place E on A, and ascertain how far they will descend between the first and third strokes of the bell, *i.e.* in double the time employed before. It will be found that they will descend in the two units of time about four times as far as during the first unit of time. Later on it will be shown that, in order to accomplish this, the average velocity during the second unit of time must be twice that during the first unit of time. If MV represent the momentum generated during the first unit of time, then the momentum generated during the second unit of time must be about $2MV$.

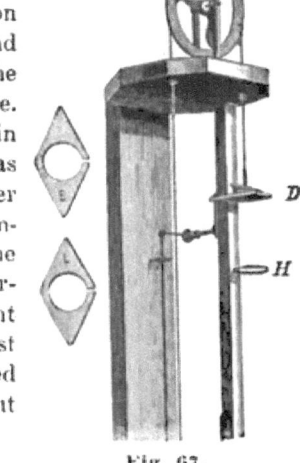

Fig. 67.

(2) *The momentum generated by a given force is proportional to the time during which the force acts.*

Section II.

FIRST LAW OF MOTION.

The relations between matter and force are concisely expressed in what are known as *The Three Laws of Motion* first enunciated by Sir Isaac Newton.

61. First Law of Motion. — *A body at rest remains at rest, and a body in motion moves with uniform velocity in a straight line, unless acted upon by some external force.*

That part of the law which pertains to motion is briefly

summarized in the familiar expression, "perpetual motion." "Is perpetual motion possible?" has been often asked. The answer is simple,—Yes, more than possible, *necessary, if no force interferes to prevent.* What has a person to do who would establish perpetual motion? Isolate a moving body from interference of all external forces, such as gravity, friction, and resistance of the air. *Can the condition be fulfilled?*

In consequence of its utter inability to put itself in motion or to stop itself, every body of matter tends to remain in the state that it is in with reference to motion or rest; this inability is called *inertia*. The First Law of Motion is often appropriately called the Law of Inertia.

Section III.

SECOND LAW OF MOTION.

62. Graphical Representation of Motion and Force. — If a person wishes to describe to you the motion of a ball struck by a bat, he must tell you three things: (1) *where it starts*, (2) *in what direction it moves*, and (3) *how far it goes*. These three essential elements may

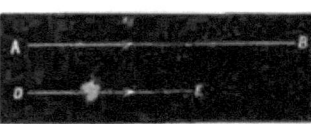

Fig. 68.

be represented graphically by lines. Thus, suppose balls at A and D (Fig. 68) to be struck by bats, and that they move respectively to B and E in one second. Then the points A and D are their starting-points; the lines AB and DE represent the direction of their motions, and the lengths of the lines represent the distances traversed. In reading, the direction should be indicated by the order of the letters, as AB and DE.

Likewise, the *forces* which produce the motion may be represented graphically. For example, the points A and B may represent the points where the forces begin to act, the lines AB and DE represent the direction in which they act, and the length of the lines represent their relative intensities.

Let a force whose intensity may be represented numerically by 8 (*e.g.* 8 pounds), acting in the direction AB (Fig. 69), be applied continuously to a ball starting at A, and suppose this force capable of moving it to B in one second; now, at the end of the second let a force of the intensity of 4, directed at right angles to the direction of the former force, act during a second — it would move the ball to C. If, however, when the ball is at A, *both* of these forces should be applied *at the same time*, then at the end of a second the ball will be found at C. Its path will not be AB nor AD, but an intermediate one, AC. Still each force produces its own peculiar result, for neither alone would carry it to C, but both are required.

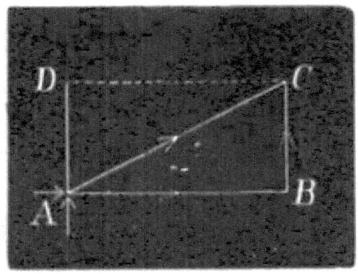

Fig. 69.

63. Second Law of Motion. — *Change of momentum is in the direction in which the force acts, and is proportional to its intensity and the time during which it acts.*

This law implies that *an unbalanced force of the same intensity, in the same time, always produces exactly the same change of momentum, regardless of the mass of the body on which it acts, and regardless of whether the body is in motion or at rest, and whether the force acts alone or with others at the same time.*

Section IV.

COMPOSITION AND RESOLUTION OF FORCES.

64. Composition of Forces. — It is evident that a single force, applied in the direction AC (Fig. 69), might produce the same result that is produced by the two forces represented by AB and AD. Such a force is called a *resultant*. *A resultant is a single force that may be substituted for two or more forces, and produce the same result that the simultaneous action of the combined forces produce.* The several forces that contribute to produce the resultant are called its *components*. When the components are given, and the resultant required, the problem is called *composition of forces. The resultant of two forces acting simultaneously at an angle to each other may always be represented by a diagonal of a parallelogram, of which the two adjacent sides represent the components.* Thus, the lines AD and AB represent respectively the direction and relative intensity of each component, and AC represents the direction and intensity of the resultant.

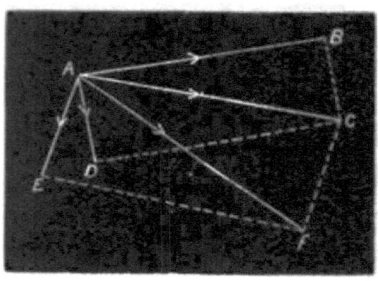

Fig. 70.

The numerical value of the resultant may be found by comparing the length of the line AC with the length of either AB or AD, whose numerical values are known. Thus, AC is 2.23 times AD; hence, the numerical value of the resultant AC is $(4 \times 2.23 =)$ 8.92.

When more than two components are given, find the result-

COMPOSITION AND RESOLUTION OF FORCES. 73

ant *of any two of them, then of this resultant and a third, and so on until every component has been used.* Thus, in Fig. 70, AC is the resultant of AB and AD, and AF is the resultant of AC and AE, *i.e.* of the three forces AB, AD, and AE. Generally speaking, a *motion may be the result of any number of forces.* When we see a body in motion, we cannot determine by its behavior how many forces have concurred to produce its motion.

65. Resolution of Forces. — Assume that a ball moves a certain distance in a certain direction, AC (Fig. 71), under the combined influence of two forces, and that one of the forces that produces this motion is represented in intensity and direction by the line AB: what must be the intensity and direction of the other force? Since AC is the resultant of two forces acting at an angle to each other, it is the diagonal of a parallelogram of which AB is one of the sides. From C draw CD parallel with and equal to BA, and complete the parallelogram by connecting the points B and C, and A and D. Then, according to the principle of composition of forces, AD represents the intensity and direction of the force which, combined with the force AB, would move the ball from A to C. The component AB being given, no other single force than AD will satisfy the question.

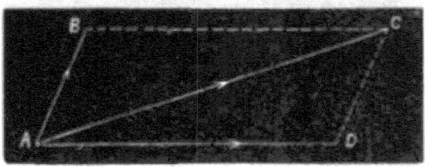

Fig. 71.

Experiment 62. — Verify the preceding propositions in the following manner: From pegs A and B (Fig. 72), in the frame of a blackboard, suspend a known weight W, of (say) 10 pounds, by means of two strings connected at C. In each of these strings insert dynamometers x and y. Trace upon the blackboard short lines along the strings from the point C, to indicate the direction of the two component forces; also trace the line CD, in continuation of the line WC, to indicate the direction and intensity of the resultant. Remove

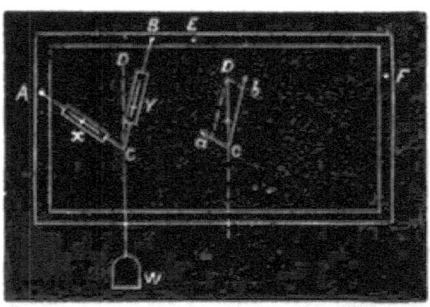

Fig. 72.

the dynamometers, extend the lines (as C*a* and C*b*), and on these construct a parallelogram, from the extremities of the line CD regarded as a diagonal. It will be found that 10 : number of pounds indicated by the dynamometer *x* : : CD : C*a*; also that 10 : number of pounds indicated by the dynamometer *y* : : CD : C*b*. Again, it is plain that a single force of 10 pounds must act in the direction CD to produce the same result that is produced by the two components. Hence, *when two sides of a parallelogram represent the intensity and direction of two component forces, the diagonal represents the resultant.* Vary the problem by suspending the strings from different points, as E and F, A and F, etc.

An excellent verification of the Second Law of Motion and the principle of composition of forces is found in the fact that a ball, projected horizontally, will reach the ground in precisely the same time that it would if dropped from a state of rest from the same hight. That is, any previous motion a body has in any direction does not affect the action of gravity upon the body.

Experiment 63. — Draw back the rod *d* (Fig. 73) toward the left, and place the detent-pin *c* in one of the slots. Place one of the brass balls on the projecting rod, and in contact with the end of the instrument, as at A. Place the other ball in the short tube B. Raise the apparatus to as great an elevation as practicable, and place it in a perfectly horizontal position. Release the detent, and the rod, propelled by the elastic force of the spring within, will strike the ball B with great force, projecting it in a horizontal direction. At the same instant that B leaves the tube and is free to fall, the ball A is released from the rod, and begins to fall. The sounds made on striking the floor reach the ears of the observer at the same instant; this shows that both balls reach the floor in sensibly the same time, and that the horizontal motion which one of the balls has does not affect the time of its fall.

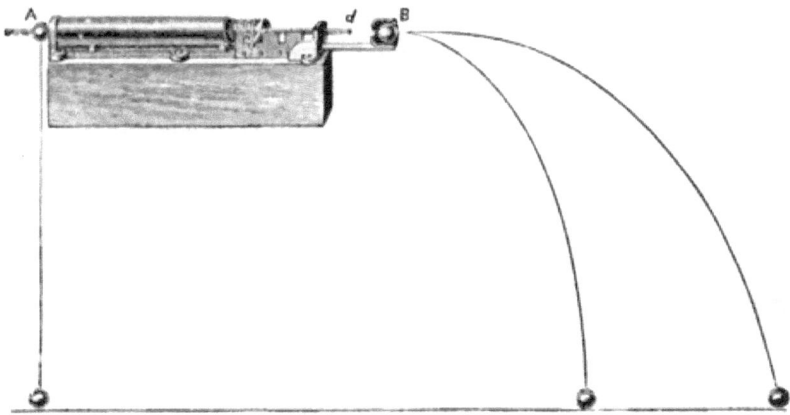

Fig. 73.

66. Composition of Parallel Forces. — If the strings CA and CB (Fig. 72) are brought nearer to each other (as when suspended from B and E) so that the angle formed by them is diminished, the component forces, as indicated by the dynamometers, will decrease, till the two forces become parallel, when the sum of the components just equals the weight W. Hence, (1) *two or more forces applied to a body act to the greatest advantage when they are parallel, and in the same direction, in which case their resultant equals their sum.*

On the other hand, if the strings are separated from each other, so as to increase the angle formed by them, the forces necessary to support the weight increase until they become exactly opposite each other, when the two forces neutralize each other, and none is exerted in an upward direction to support the weight. If the two strings are attached to opposite sides of the weight (the weight being supported by a third string), and pulled with equal force, the weight does not move. But if one is pulled with a force of 15 pounds, and the other with a force of 10 pounds, the weight moves in the direction of the greater force; and if a third dynamometer is attached to the weight, on the side of the weaker force, it is found that an additional force of five pounds must be applied to prevent motion. Hence, (2) *when two or more forces are applied to a body, they act to greater disadvantage the farther their directions are removed from one another; and the result of parallel forces acting in opposite directions is a resultant force in the direction of the greater force, equal to their difference.*

When parallel forces are not applied at the same point, the question arises, What will be the point of application of their resultant? To the opposite extremities of a bar AB (Fig. 74) apply two sets of weights, which shall be to each other as 3 lbs. : 1 lb. The resultant is a single force, applied at some point between A and B.

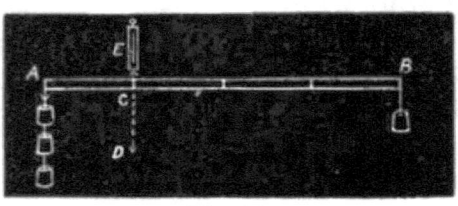

Fig. 74.

To find this point it is only necessary to find a point where a single force, applied in an opposite direction, will prevent motion resulting from the parallel forces; in other words, to find a point where a support may be applied so that the whole will be balanced. That point is found by trial to be at the point C, which divides the bar into two parts so that AC : CB :: 1 lb. : 3 lbs. Hence, (3) *when two parallel forces act upon a body in the same direction, the distances of their points of applica-*

tion *from the point of application of their resultant are inversely as their intensities.*

The dynamometer E indicates that a force equal to the sum of the two sets of weights is necessary to balance the two forces. A force whose effect is to balance the effects of several components is called an *equilibrant*. The resultant of the two components is a single force, equal to their sum, applied at C in the direction CD.

67. Moment of a Force. — The tendency of a force to produce rotation about a fixed point as C (Fig. 75) is called its *moment* about that point. The perpendicular distance (AC or BC) from the fixed point (C) to the

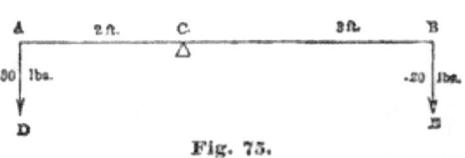

Fig. 75.

line of direction in which the force acts (AD or BE) is called the *leverage* or *arm*. *The moment of a force is measured by the product of the number of units of force into the number of units of leverage.* For example, the moment of the force applied at A is expressed numerically by the number $(30 \times 2 =)$ 60.

68. Equilibrium of Moments. — The moment of a force is said to be *positive* when it tends to produce rotation in the direction in which the hands of a clock move, and *negative* when its tendency is in the reverse direction. If two forces act at different points of a body which is free to rotate about a fixed point, they will produce equilibrium when their moments are opposite and their algebraic sum is zero. Thus the moment of the force applied at A (Fig. 75) is $(-30 \times 2) - 60$. The moment of the force applied at B in an opposite direction is accordingly $(+20 \times 3 =) + 60$. Their algebraic sum is zero, consequently there is equilibrium between the forces.

When more than two forces act in this manner, there will be equilibrium if the sum of all the positive mo-

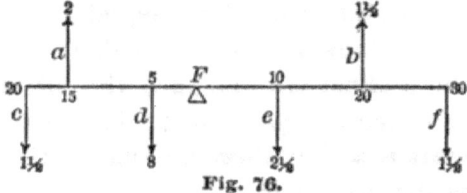

Fig. 76.

ments is equal to the sum of all the negative moments. Thus the sum of the positive moments acting about point F (Fig. 76) is $(f) 45 + (e) 25 + (a) 30 = 100$; the sum of the negative moments acting about the same point is $(c) 30 + (d) 40 + (b) 30 = 100$; the two sums being equal, the forces are in equilibrium.

69. Mechanical Couple. — If two equal, parallel, and contrary forces are applied to opposite extremities of a stick AB (Fig. 77), no single force can be applied so as to keep the stick from moving; there

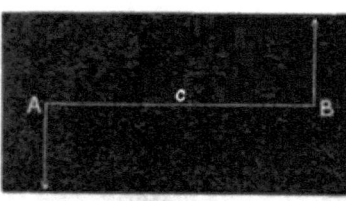

Fig. 77.

will be no motion of translation, but simply a *rotation* around its middle point C. Such a pair of forces, equal, parallel, and opposite, is called a *mechanical couple*.

Section V.

THE THIRD LAW OF MOTION.

70. Introductory Experiments. — We have learned that motion cannot originate in a single body, but arises from mutual action between two bodies or two parts of a body. For example, a man can lift himself by pulling on a rope attached to some other object, but not by his boot-straps, or a rope attached to his feet. *In every change in regard to motion there are always at least two bodies oppositely affected.*

Experiment 64. — Suspend the deep glass bucket A (Fig. 78) by means of a strong thread two feet long, so that the long projecting pointer may be directly over a dot made on a piece of paper placed beneath; or place beneath another pointer, B, so that the two points shall just meet. Fill the bucket with water. Gravity causes the water to flow from the orifice C; but *the bucket moves in the opposite direction.*

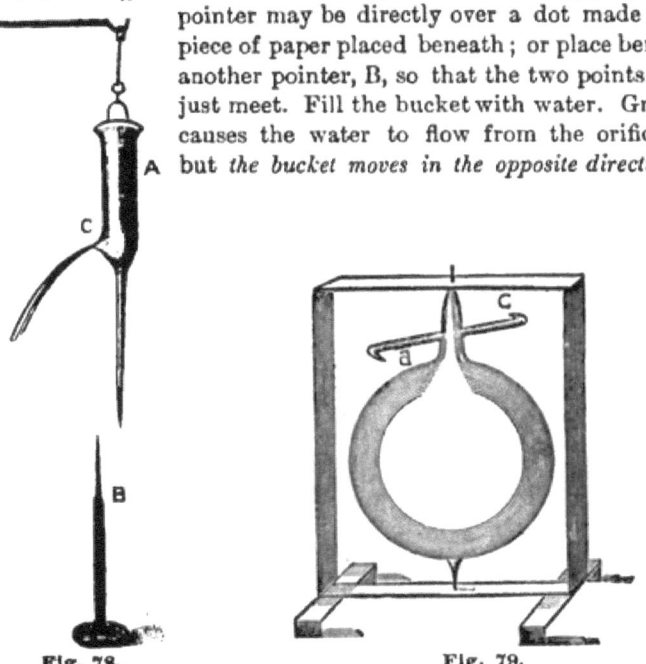

Fig. 78.   Fig. 79.

Experiment 65. — Place the hollow glass globe and stand (Fig. 79) under the receiver of an air-pump, and exhaust the air. The air within the globe expands, and escapes from the small orifices a and c at the extremity of the two arms. But this motion of the air is attended by an opposite motion of the arms and globe, and a rapid rotation is caused.

A man in a boat weighing one ton pulls at one end of a rope, the other end of which is held by another man, who weighs twice as much as the first man, in a boat weighing two tons: both boats will move towards each other, but in opposite directions; if the resistances which the two boats encounter were the same, the lighter boat would move twice as fast as the heavier, *but with the same momentum*.

If the boats are near each other, and the men push each other's boats with oars, the boats will move in opposite directions, though with different velocities, yet with equal momenta.

The opposite impulses received by the bodies concerned are usually distinguished by the terms *action* and *reaction*. We measure these, when both are free to move, by the momenta generated, which is always the same in both bodies.

71. Third Law of Motion. — *To every action there is an equal and opposite reaction.*

The application of this law is not always obvious. Thus, the apple falls to the ground in consequence of the mutual attraction between the apple and the earth. The earth does not appear to fall toward the apple. But, as the mass of the earth is enormously greater than that of the apple, its velocity, for an equal momentum, is proportionately less.

EXERCISES.

1. (*a*) Why does not a given force, acting the same length of time, give a loaded car as great a velocity as an empty car? (*b*) After equal forces have acted for the same length of time upon both cars, and given them unequal velocities, which will be the more difficult to stop?

2. (*a*) The planets move unceasingly; is this evidence that there are forces pushing or pulling them along? (*b*) None of their motions are in straight lines; are they acted upon by external forces?

THE THIRD LAW OF MOTION.

3. A certain body is in motion; suppose that all hindrances to motion and all external forces were withdrawn from it, how long would it move? Why? In what direction? Why? With what kind of motion, *i.e.* accelerated, retarded, or uniform? Why?

4. Copy upon paper and find the resultant of the components AB and AC in each of the four diagrams in Figure 80. Also assign appropriate numerical values to each component, and find the corresponding numerical value of each resultant.

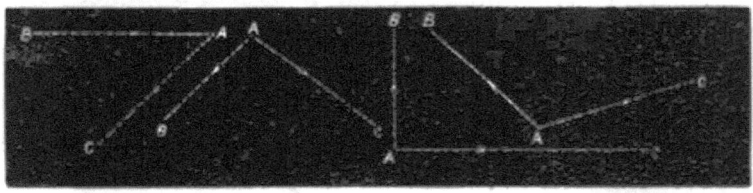

Fig. 80.

5. Explain how rotating lawn-sprinklers are kept in motion.

6. When you leap from the earth, which receives the greater momentum, your body or the earth?

7. When you kick a door-rock, why does snow or mud on your shoes fly off?

8. Why cannot a person propel a vessel during a calm by blowing the sails with a big bellows placed on the deck of the same vessel?

9. In swimming, you put water in motion; what causes your body to advance? What propels the bird in flying?

10. Could a rocket be projected in the usual way if there were no atmosphere?

Section VI.

APPLICATIONS OF THE THREE LAWS OF MOTION. — CENTER OF GRAVITY.

72. Center of Gravity Defined. — Let Figure 81 represent any body of matter; for instance, a stone. Every molecule of the body is acted upon by the force of gravity.

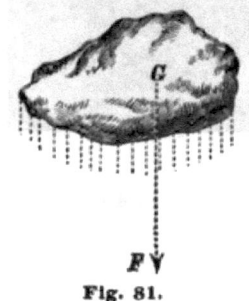

Fig. 81.

The forces of gravity of all the molecules form a set of parallel forces acting vertically downward, the resultant of which equals their sum, and has the same direction as its components. The resultant passes through a definite point in whatever position the body may be, and this point is called its *center of gravity*. *The center of gravity (c.g.) of a body is, therefore, the point of application of the resultant of all these forces;* and for practical purposes *the whole weight of the body may be supposed to be concentrated at its center of gravity.*

Let G in the figure represent this point. For practical purposes, then, we may consider that gravity acts only upon this point, and in the direction GF. If the stone falls freely, this point cannot, in obedience to the first law of motion, deviate from a vertical path, however much the body may rotate about this point during its fall. Inasmuch, then, as the c.g. of a falling body always describes a definite path, a line GF that represents this path, or the path in which a body supported tends to move, is called the *line of direction*.

It is evident that if a force is applied to a body equal to

its own weight, and opposite in direction, and anywhere in the line of direction (or its continuation), this force will be the *equilibrant* of the forces of gravity; in other words, the body subjected to such a force is in equilibrium, and is said to be *supported*, and *the equilibrant* is called its *supporting force.* *To support any body*, then, *it is only necessary to provide a support for its center of gravity. The supporting force must be applied somewhere in the line of direction, otherwise the body will fall.* The difficulty of poising a book, or any other object, on the end of a finger, consists in keeping the support under the center of gravity.

Figure 82 represents a toy called a "witch," consisting of a cylinder of pith terminating in a hemisphere of lead. The toy will not lie in a horizontal position, as shown in the figure, because the support is not applied immediately under its c.g. at G; but when placed horizontally, it immediately assumes a vertical position. It appears to the observer to rise; but, regarded in a mechanical sense, it really falls, because its c.g., where all the weight is supposed to be concentrated, takes a lower position.

Fig. 82.

73. How to Find the Center of Gravity of a Body. — Imagine a string to be attached to a potato by means of a tack, as in Figure 83, and to be suspended from the hand. When the potato is at rest, there is an equilibrium of forces, and the c.g. must be somewhere in the line of direction *an*; hence, if a knitting-needle is thrust vertically through the potato from *a*, so as to represent a continuation of the vertical line *oa*, the c.g. must lie somewhere in the

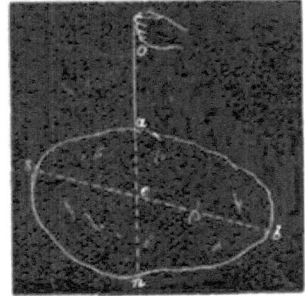

Fig. 83.

path *an* made by the needle. Suspend the potato from some other point, as *b*, and a needle thrust vertically through the potato from *b* will also pass through the c.g. Since the c.g. lies in both the lines *an* and *bs*, it must be at *c*, their point of intersection. It will be found that, from whatever point the potato is supported, the point *c* will always be vertically under the point of support. On the same principle the c.g. of any body is found. But the c.g. of a body may not be coincident with any particle of the body; for example, the c.g. of a ring, a hollow sphere, etc.

74. Equilibrium of Bodies. — That a body acted on solely by its weight may be in equilibrium (*i.e.* supported), it is sufficient that its line of direction shall pass through the point or surface by which it is supported. For example, when a body is to be supported at its base, the line of direction must pass through the base. The base of a body is not necessarily limited to that part of the under surface of a body that touches its support. For example, if a string is placed around the four legs of a table near the floor, the rectangular figure bounded by the string is the base of the table.

It is evident that the resultant weight of a body acting at its c.g. tends to bring this point as low as possible; hence *a body tends to assume a position such that its c.g. will be as low as possible.*

In whatever manner a body is supported, the equilibrium is *stable* if, on moving the body, the center of gravity ascends; *unstable*, if it descends; and *neutral*, if it neither ascends nor descends, as that of a sphere rolled on a horizontal plane.

Experiment 66. — Try to support a ring on the end of a stick, as at *b* (Fig. 84). If you can keep the support exactly under the c.g. of

APPLICATIONS OF THE THREE LAWS OF MOTION. 85

the ring, there will be an equilibrium of forces, and the ring will remain at rest. But if it is slightly disturbed, the equilibrium will be destroyed, and the ring will fall. Support it at *a*; in this position its c.g. is as low as possible, and any disturbance will raise its c.g.; but, in consequence of the tendency of the c.g. to get as low as possible, it will quickly fall back into its original position.

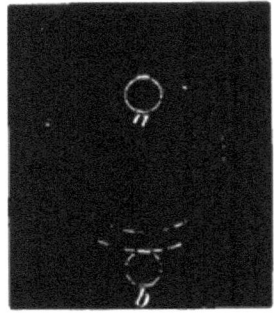

Fig. 84.

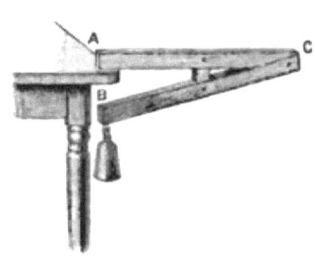

Fig. 85.

Experiment 67.— Prepare a V-shaped frame like that shown in Figure 85, the bar AC being about three feet long; place it so that the end will overlap the table two or three inches, and hang a heavy weight or a pail of water on the hook B, and the whole will be supported. Rock the weight back and forth by raising the end C and allowing it to fall. What kind of equilibrium is this? Remove the weight, and the bar falls to the floor. Why?

The stability of a body varies with its breadth of base, and inversely with the hight of its c.g. above its base. Support a book on a table so that it may have three different degrees of stability, and account for the same.

QUESTIONS.

1. Why is a person's position more stable when his feet are separated a little, than when close together?
2. How does ballast tend to keep a vessel from overturning?
3. For what two reasons is a pyramid a very stable structure?
4. What point in a falling body descends in a straight

Fig. 86.

line? What is this line called? Disregarding the motions of the earth, toward what point in the earth does this line tend?

5. It is difficult to balance a lead-pencil on the end of a finger; but by attaching two knives to it, as in Figure 86, it may be rocked to and fro without falling. Explain.

Section VII.

APPLICATIONS OF THE THREE LAWS OF MOTION CONTINUED. — EFFECT OF A CONSTANT FORCE ACTING ON A BODY PERFECTLY FREE TO MOVE. — FALLING BODIES.

75. Any Force, however Small, can move any Body of however Great Mass. — For example, a child can move a body having a mass equal to that of the earth, provided only that the motion of this body is not hindered by a third body. Moreover, the amount of momentum that the child can generate in this immense body in a given time is precisely the same as that which it would generate by the exertion of the same force for the same length of time on a body having a mass of (say) 10 pounds. Momentum is the product of mass into velocity; so, of course, as the mass is large, the velocity acquired in a given time will be correspondingly small. The instant the child begins to act, the immense body begins to move. Its velocity, infinitesimally small at the beginning, would increase at almost an infinitesimally slow rate, so that it might be months or years before its motion would become perceptible. It is easy to see how persons may get the impression that very large bodies are immovable except by very great forces. The erroneous idea is acquired that bodies of matter have a power to resist the action of forces in causing motion, and that the greater the mass, the greater the resistance ("quality of not yielding to force," *Webster*). The fact is, *that no body of whatever mass has any power to resist motion;* in other words, "*a body free to move cannot remain at rest under the slightest unbalanced force tending to set it in motion.*" Furthermore, *a given force acting for the same length of time will generate the same amount of momentum in all bodies free to move, irrespective of their masses.*

76. Falling Bodies. — A constant force is one that acts continuously and with uniform intensity. Nature furnishes no example of a body moved by a force so nearly constant as that of a body falling through a moderate distance to the earth. Inasmuch as the velocity of falling bodies is so great that there is not time for accurate observation during their fall, we must, in investigating the laws of falling bodies, resort to some method of checking their velocity, without otherwise changing the character of the fall.

Experiment 68. — Ascertain, as in Experiment 60, how far the weights, moved by a constant force (*e.g.* 2 grams), descend during one swing of the pendulum. Inasmuch as all swings of the pendulum are made in equal intervals of time, we may take the time of one swing as our *unit of time*. We will, for convenience, take for our *unit of distance* the distance the weights fall during the first unit of time, call this unit a *space*, and represent the unit graphically by the line *ab* (Fig. 87).

Next ascertain how far the weights fall from the starting-point during two units of time (*i.e.* two swings of the pendulum). The distance will be found to be four spaces, or four times the distance that they fell during the first unit of time. This distance is represented by the line *ac*. But we have learned that the weights descend only one space (*ab*) during the first unit of time, hence they must descend three spaces during the second unit of time. The weights, under the action of the constant force, start from a state of rest, and move through one space in a unit of time. This force, continuing to act, accomplishes no more nor less during any subsequent unit of time. But the weights move through three spaces during the second unit of time; hence two of the spaces must be due to the velocity they had acquired at the end of the first unit. In other words, if the ring H is placed at the point (corresponding to *b*) reached by the weights at the end of the first unit of time, then weight E will be caught off (*i.e.* the constant force will be withdrawn), and the other weights will, in conformity with the first law of motion, continue to move with uniform velocity from this point (except as they are retarded by resistance of the air and the friction of the wheel C), and will descend two spaces during the second unit and reach point *e*. (Try it.)

The weights, therefore, have at the end of the first unit of time a velocity (V) of two spaces. But they started from a state of rest: hence the constant force causes, during the first unit of time, an acceleration of velocity equal to two spaces.

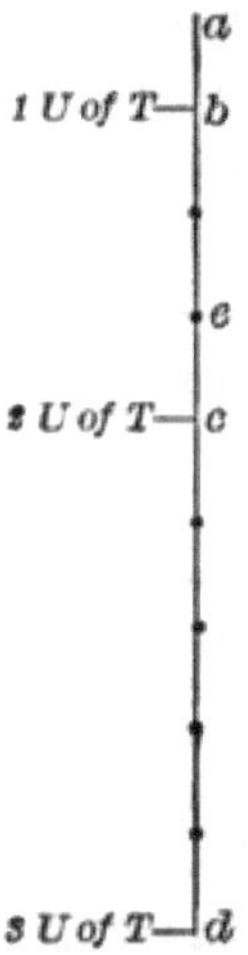

Fig. 87.

Let the weights descend three units of time, and it will be found that the weights will descend in this time nine spaces (*ad*), or five spaces (*cd*) during the third unit of time. One of these five spaces is due to the action of the force during the third unit of time; the weights must then have had at point *c* (*i.e.* at the end of the second unit of time) a velocity of four spaces. But at the end of the first unit of time they had a velocity of two spaces; then they must have gained during the second unit of time a velocity of two spaces. It seems, then, that *the effect of a constant force applied to a body is to produce uniformly accelerated motion* when there are no resistances.

The acceleration due to gravity is usually represented by *g*, and is always twice the distance ($\frac{1}{2} g$) traversed during the first unit of time. When a body is acted upon by any other constant force, the acceleration produced by the force is usually represented by the letter A.

APPLICATIONS OF THE THREE LAWS OF MOTION.

Arrange the results of your observations in a tabulated form as follows: —

No. of units of time.	Total distance passed over. (S)	Distance passed over in each unit; also average velocity. (s)	Velocity at the end of each unit. (V)	Increase of velocity in each unit, i.e. acceleration.
1	1 ($\frac{1}{2}g$)	1 ($\frac{1}{2}g$)	2 ($\frac{1}{2}g$)	2 ($\frac{1}{2}g$)
2	4 "	3 "	4 "	2 "
3	9 "	5 "	6 "	2 "
4	16 "	7 "	8 "	2 "
etc.	etc.	etc.	etc.	etc.

77. Formulas for Uniformly Accelerated Motion. — If we substitute A for g, and represent the distance traversed during a given unit of time by s, and the total distance the body has accomplished from the outset to the end of a given unit of time (T) by S, we derive from our tabulated results the following formulas for solving problems of uniformly accelerated motion: —

(1) $V = (\frac{1}{2} A \times 2 T) = A T$.
(2) $s = \frac{1}{2} A (2 T - 1)$.
(3) $S = \frac{1}{2} A T^2$.

Hence, (1) *the velocity acquired varies as the time;* (2) *the spaces passed over in successive equal intervals of time vary as the odd numbers* 1, 3, 5, 7, etc.; and (3) *the entire space traversed varies as the square of the time.*

Strictly speaking, a falling body is not under the influence of a constant force, inasmuch as gravity varies inversely as the square of the distance from the center of the earth. But for small distances the variation may be, for all practical purposes, disregarded, as at a hight of a kilometer (about $\frac{2}{3}$ of a mile) it is only about $\frac{1}{3188}$ of the weight at the surface. It can be shown mathematically that the velocity that would be acquired by a body falling freely to the earth's surface from an infinite distance would be about 35,000 feet per second.

78. Velocity of a Falling Body Independent of its Mass and Kind of Matter. — If we grasp a coin and a bit of paper between the thumb and finger, and release both at the same instant, the coin will reach the floor first. It would seem as though a heavy body falls faster than a light body. Galileo was the first to show the falsity of this assumption. He let drop from an eminence iron balls of different weights: they all reached the ground at the same instant. Hence he concluded that *the velocity of a falling body is independent of its mass.*

He also dropped balls of wax with the iron balls. The iron balls reached the ground first. Are some kinds of matter affected more strongly by gravitation than others? If a coin and several bits of paper are placed in a long glass tube (Fig. 88), the air exhausted, and the tube turned end for end, it will be found that the coin and the paper will fall with equal velocities. Hence, *the earth attracts all matter alike.* A wax ball of the same size as an iron ball meets with the same resistance from the air that the iron ball does; but since the mass of the former is less than that of the latter, the force acting on the former is less, and a less force cannot overcome the same resistance as quickly, consequently in the air the wax ball falls a little more slowly. We conclude, therefore, that *in a vacuum all bodies fall with equal velocities.*

Fig. 88.

Experiments show that in the latitude of the Northern States the acceleration, *i.e.* the value of g, is, near sea-level and in a vacuum, $32\frac{1}{6}$ feet (9.8^m) per second; that is, the velocity gained by a falling body, disregarding the resistance of the air, is $32\frac{1}{6}$ feet per second, and the body falls in the first second $16\frac{1}{12}$ feet (4.9^m).

EXERCISES.

1. What is a constant force? What effect does it produce on every body when there are no resistances?

2. (*a*) How far will a body fall in a vacuum in one second? (*b*) What is its velocity at the end of the first second? (*c*) What is its acceleration per second?

3. (*a*) How far will a body fall in ten seconds? (*b*) How far will it fall in the tenth second? (*c*) What is its velocity at the end of the tenth second? (*d*) What is its average velocity during the tenth second?

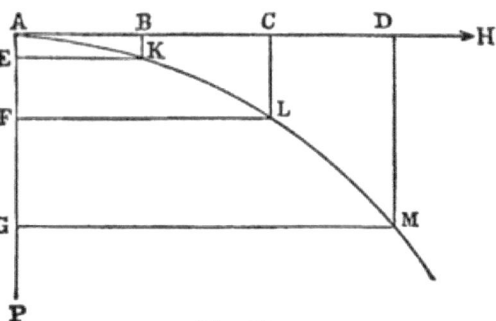

Fig. 89.

4. (*a*) How far will a body fall in one-fourth of a second? What is the velocity of a falling body at the end of the first quarter of a second of its fall?

5. A body is projected from point A (Fig. 89) in the horizontal direction AH. (*a*) If there were no resistance of the air, and gravity did not act on it, it would go a distance during the first unit of time represented by AB; how far would it go during the second and third units of time? (In every answer quote the law of motion in conformity with which your answer is given.) (*b*) If the body were dropped from A, it would reach successively points E, F, and G at the ends of the first, second, and third units of time. If the body were projected horizontally in the direction AH, and gravity acts during its flight, what points will the body successively reach at the end of the same units of time?

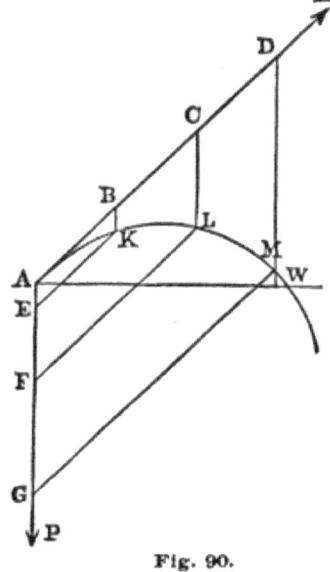

Fig. 90.

6. (*a*) Suppose that a body is projected obliquely upward in the direction AH (Fig. 90), (gravity meantime acting on the body); what points will the body reach successively at the end of the first, second, and third units of time? (*b*) How far will the ascending body virtually fall during the first unit of time? (*c*) How far during the second unit? (*d*) How far during the third unit? (*e*) Show that your answers are consistent with the Second Law of Motion.

7. (*a*) Under the action of a constant force, a body meeting with no resistances moves from a state of rest 20 feet during the first minute: how far will it go in an hour? (*b*) Suppose at the end of the first minute the force should cease to act, how far would the body go in an hour from that instant?

Section VIII.

APPLICATIONS OF THE THREE LAWS OF MOTION CONTINUED. — CURVILINEAR MOTION.

79. How Curvilinear Motion is Produced. — Motion is *curvilinear* when its direction changes at every point. But according to the first law of motion, every moving body proceeds in a straight line, unless compelled to depart from it by some external force. Hence curvilinear motion can be produced only by an external force acting continuously upon the body at an angle to the straight path in which the body tends to move, so as constantly to change its direction. In case the body moves in a circle, this force acts at right angles to the path of the body or towards the center of motion; hence this deflecting force has received the name of *centripetal force*.

80. Centrifugal Force.

Experiment 69. — Cause a ball to rotate around your hand by means of a string attached to it and held in the hand. Observe closely every phase of the operation. First, you make a movement as if to project the ball in a straight line. Immediately you begin to pull on the string to prevent its going in a straight line. By a continuous exertion of these two forces in a short time the ball acquires great speed. You may now cease to exert any projecting force, and simply keep the hand still; but as the ball has acquired a motion, and all motion tends to be in a straight line, you are still obliged to exert a pulling force to deflect it from this path. Observe that as the velocity of the ball is retarded by the resistance of the air, the pulling or deflecting force which you are obliged to employ rapidly diminishes.

To satisfy yourself that the ball tends to move in a straight line, let go the string or cut it, and the ball immediately moves off in a straight line, or simply perseveres in the direction it had at the instant the string was cut. Observe that the ball appears while rotating to be pulling your hand; but you know that all the force concerned originates in yourself, and that this apparent pull on the part of the ball is only the effect of the *reaction* of the force which you exert on the ball. This apparent reactionary force is called *centrifugal force*.

Centrifugal force is the reaction of a revolving body on the body that guides it, and is equal and opposite to the centripetal force (see Third Law of Motion).

When you swing the ball about your hand you discover that the force of the pull increases with the velocity, and more rapidly than the velocity. Careful observations have determined that for bodies revolving in circular orbits *the centripetal* (and, of course, centrifugal) *force varies as the mass of the body and the square of its velocity.*

The farther a point is from the axis[1] of motion of a rigid body, the farther it has to move during a rotation; consequently the greater its velocity. Hence, bodies situated at the earth's equator have the greatest velocity, due to the earth's rotation, and consequently the greatest tendency to fly off from the surface, the effect of which is to neutralize, in some measure, the force of gravity. It is calculated that a body weighs about $\frac{1}{289}$ less at the equator than at either pole, in consequence of the greater centrifugal force at the former place. But 289 is the square of 17; hence,

[1] *Axis*: an imaginary straight line passing through a body about which it rotates.

If the earth's velocity were increased seventeen-fold, objects at the equator would weigh nothing.

We have also learned (page 17) that a body weighs more at the poles, in consequence of the oblateness of the earth. This is estimated to make a difference of about $\frac{1}{568}$. Hence a body will weigh at the equator $\frac{1}{565}+\frac{1}{289}=$ (about) $\frac{1}{192}$ less than at the poles.

The attraction between the sun and the earth causes these bodies to move in curvilinear paths, performing what is called annual revolutions. The motion of both these bodies, were it not for this mutual attraction (and the attraction of other celestial bodies),

Fig. 91.

would be eternally in straight lines, but in consequence of their mutual attraction both rotate about a point C (Fig. 91), which is the center of gravity of the two bodies considered as one body (as if connected by a rigid rod). If both bodies had equal masses, the center of gravity and center of motion would be half-way between the two bodies; but as the mass of the earth is less than that of the sun, so its velocity and distance traversed are proportionally greater.

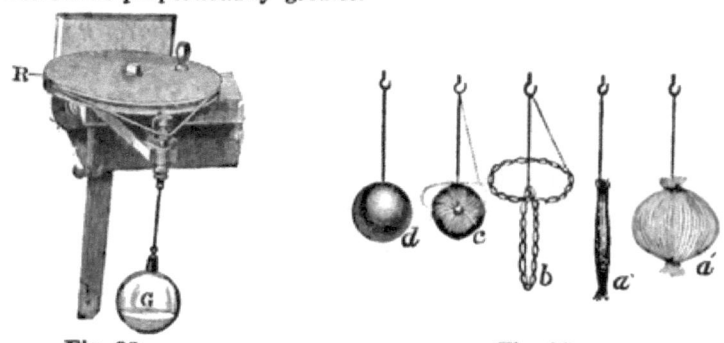

Fig. 92.   Fig. 93.

Experiment 70. — Arrange some kind of rotating apparatus, *e.g.* R (Fig. 92). Suspend a skein of thread *a* (Fig. 93) by a string, and rotate; it assumes the shape of the oblate spheroid *a'*. Suspend a glass globe G (Fig. 92) about one-tenth full of colored water, and rotate. The liquid gradually leaves the bottom, rises, and forms an equatorial ring within the glass. This illustrates the probable method by which the earth, on the supposition that it was once in a fluid state, assumed its present spheroidal state. (Explain.) Pass a string through the longest diameter of an onion *c*, and rotate; the onion gradually changes its position so as to rotate on its shortest axis.

It may be demonstrated mathematically, as well as experimentally, that *a freely rotating body is in stable equilibrium only when rotating about its shortest diameter;* hence the tendency of a rotating body to take this position.

QUESTIONS.

1. (*a*) What is the cause of the stretching **force** exerted on the rubber cord when you swing a return-ball about your hand? (*b*) Suppose that you double the velocity of the ball; how many times will you increase this stretching force?

2. Why do wheels and grindstones, when **rapidly rotating**, tend to break, and the pieces fly off?

3. On what does the magnitude of the pull between a rotating body and its center of motion depend?

4. (*a*) Explain the danger of a carriage being **overturned** in turning a corner. (*b*) How many fold is the tendency to overturn increased by doubling the velocity of the carriage?

Section IX.

APPLICATION OF THE THREE LAWS OF MOTION CONTINUED. — THE PENDULUM.

81. Laws of the Pendulum.

Experiment 71. — Suspend iron balls by strings, as in Figure 94. Make A and B the same length. Draw A and B one side, and to different hights, so that one may swing through a longer arc than the other, and let both drop at the same instant. One moves much faster than the other, and completes a longer journey at each swing, but both complete their swing or vibration at the same time.

Hence (1) *the time of vibration of a pendulum is* (strictly speaking, approximately) *independent of the length of the arc.*

Experiment 72. — Set all the balls swinging; only A and B swing together, *i.e.* in the same time. The shorter the pendulum, the faster it swings. Make B about 39 inches long from the point of suspension to the center of the ball, regulating this length, as necessity may require, so that the number of vibrations made by the pendulum in one minute shall be exactly 60; in other words, so that it shall "beat seconds." (Accurately, a pendulum that beats seconds is 39.09 inches long.) Make C one-fourth as long as B. Count the vibrations made by C in one minute; it makes 120 vibrations in the same time that B makes 60 vibrations. Make D one-ninth the length of B; the former makes three vibrations while the latter makes one. Consequently the time of vibration of the former is one-third that of the latter.

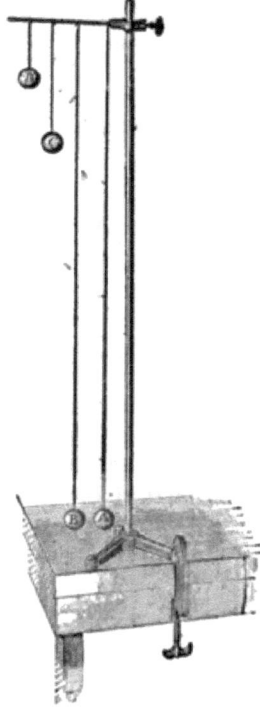

Fig. 94.

Hence (2) *the time of vibration of a pendulum varies as the square root of its length.*

By experiments too difficult for ordinary school work, it has been ascertained that (3) *the time of vibration of a pendulum varies inversely as the square root of the force of gravity* (upon which the value of g depends). Hence it is apparent that by determining the time of vibration of a pendulum of the same length, at different distances from the center of gravity of the earth (*e.g.* at the top and bottom of a mountain, or at sea-level at different latitudes), the relative value of g at these places may be ascertained.

Experiment 73. — Loosen the binding-screw in the bob of the pendulum of the Atwood machine (Fig. 66), and place the bob at different elevations on the pendulum-rod. Count the number of vibrations made by the pendulum in a minute, when the bob is placed at these different elevations. The greater the elevation of the bob, — in other words, the shorter the pendulum, — the greater the number of vibrations made. We learn by this experiment that the time of vibration of a pendulum may be regulated by raising or lowering its bob.

EXERCISES.

1. One pendulum is 20 inches long, and vibrates four times as fast as another. How long is the other?

2. (a) What effect on the rate of vibration has the weight of its bob? (b) What effect has the length of the arc? (c) What affects the rate of vibration of a pendulum?

3. How can you quicken the vibration of a pendulum threefold?

4. A clock loses time. (a) What change in the pendulum ought to be made? (b) How would you make the correction?

5. Two pendulums are four and nine feet long respectively. While the short one makes one vibration, how many will the long one make?

6. How long is a pendulum that makes two vibrations in a second?

7. What is the time of vibration of a pendulum $(39.09 \div 4 =)$ 9.75 in. long?

8. The number of vibrations made by a given pendulum in a given time varies as the square root of the force of gravity. Force of gravity at any place is expressed by the value of g (i.e. by the acceleration which it produces). (a) If at a certain place a pendulum 39.09 in. long make 3600 vibrations in an hour, and the value of g is 32.16 ft., what is the acceleration at a place where the same pendulum makes 3590 vibrations in the same time? (b) Which of the two places is nearer the center of gravity of the earth?

9. Suggest some way by which the force of gravity at different latitudes and altitudes may be determined.

10. (a) A certain body weighs 12 lbs. where the value of g is 32.16 ft.; what will the same body weigh at a place where $g = 32$ ft.? (b) Suppose that the former place is at the surface of the earth and 4000 miles from the earth's center of gravity; how far above it is the latter place? (See page 16.)

11. A pebble is suspended by a thread 2 ft. long; required the number of vibrations it will make in a minute.

12. Why do not heavy bodies fall faster than light ones in a vacuum?

13. Take equal masses of wood and lead; which weighs more?

14. A stone falls from the top of a railway carriage which is moving at the rate of one-half of a mile a minute. Find what horizontal distance and what vertical distance the stone will have passed through in one-tenth of a second, disregarding the resistance of the air.

Ans. 4.4 ft.; .16 ft.

CHAPTER IV.

WORK AND ENERGY.

Section I.

METHODS OF ESTIMATING WORK AND ENERGY.

82. Work. — *Whenever a force causes motion, it does work.* A force may act for an indefinite time without doing work; for example, a person may support a stone for a time and become weary from the continuous application of force to prevent its falling, but he does no work upon the stone because he effects no change. When a force acts through space, work is done. Let the person holding the weight exert just a little more force; the weight will rise, and work will be done.

A body that is moved is said to have work done upon it; and a body that moves another body is said to do work upon the latter. When the heavy weight of a pile-driver is raised, work is done upon it; when it descends and drives the pile into the earth, work is done upon the pile, and the pile in turn does work upon the matter in its path.

The act of doing work may consist in a mere transfer of energy from the body doing work to the body on which work is done, or it may consist in a transformation of energy from one kind to some other kind, as when the pile-driver strikes the pile and the pile is forced into the earth, a part of the energy concerned in each case is transformed into *heat*, which we shall learn farther on is *molecular energy*.

In future chapters we shall discuss the subject of transformations of energy; for the present our discussions relate chiefly to transferences of energy.

83. Formulas for Estimating Work. — Force and space (or distance) are the essential elements of work, and necessarily are the quantities employed in estimating work. A given force acting through a space of one foot, in raising a weight, does a certain amount of work; it is evident that the same force acting through a space of two feet would do twice as much work. Hence the general formula

$$FS = W, \qquad (1)$$

in which F represents the force employed, S the space through which the force acts, and W the work done.

In case a force encounters resistance, the magnitude of the force necessary to produce motion varies as the resistance. Often the work done upon a body is more conveniently determined by *multiplying the resistance by the space through which it is overcome*, and our formula becomes by substitution of R (resistance) for F (the force which overcomes it)

$$RS = W. \qquad (2)$$

For example, a ball is shot vertically upward from a rifle in a vacuum; the work done upon the ball (by the explosive force of the gunpowder) may be estimated by multiplying the average force (difficult to ascertain) exerted upon it, by the space through which the force acts (a little greater than the length of the barrel); or by multiplying the resistance to motion offered by gravity, *i.e.* its weight (easily ascertained) by the distance the ball ascends.

84. Energy, Kinetic and Potential. — Every moving body can impart motion; hence it can do work upon another body, and is therefore said to possess *energy*. The *energy of a body is its "capacity to do work."* The energy which a body possesses in consequence of its motion is called *kinetic* energy.

A stone lying on the ground is devoid of energy. Raise it and place it on a shelf; in so doing you perform work upon it. As you look at it lying motionless upon the shelf, it *appears* as devoid of energy as when lying on the earth. Attach one end of a cord to it and pass it over a pulley and wind a portion of the cord around the shaft connected with a sewing-machine, coffee-mill, lathe, or other convenient machine. Suddenly withdraw the shelf from beneath the stone. The stone moves; it communicates motion to the machinery, and you may sew, grind coffee, turn wood, etc., with the energy given to the machine by the stone.

The work done on the stone in raising it was not lost; the stone pays it back while descending. There is a very important difference between the stone lying on the floor, and the stone lying on the shelf: the former is powerless to do work; the latter can do work. Both are alike motionless, and you can see no difference, except an *advantage* that the latter has over the former in having a *position* such that it *can move*. What gave it this advantage? Work. *A body*, then, *may possess energy due merely to* ADVANTAGE OF POSITION, *derived always from work bestowed upon it.* Energy due to advantage of position is called *potential* energy. We see, then, that energy may exist in either of two widely different states. It may exist as *actual motion*, or it may exist in a *stored-up condition*, as in the stone lying on the shelf.

Possibly some will object that the work done is performed by gravity, and not by the stone; that if this force should cease to exist, the stone would not move when the shelf is removed, and consequently no work would be done. All this is very true, and it is likewise true that when the stone is on the ground, the same force of gravity is acting, but can do no work simply because the position of things is such that the stone cannot move. The energy which the stone on the shelf possesses is due to the fact that its *position* is such that it can move, and that there is a *stress* between it and the earth which will cause it to move. Both advantage of position and stress are necessary, but the former is attained only by work performed. The force of gravity is employed to do work, as when mills are driven by falling water; but the water must first be raised from the ocean-bed to the hillside by the work of the sun's heat. The elastic force of springs is employed as a motive power; but this power is due to an advantage of position which the molecules of the springs have first acquired by work done upon them.

We are as much accustomed to store up energy for future use as provisions for the winter's consumption. We store it when we wind up the spring or weight of a clock, to be doled out gradually in the movements of the machinery. We store it when we bend the bow, raise the hammer, condense air, and raise any body above the earth's surface.

A body possesses potential energy when, in virtue of work done upon it, it occupies a position of advantage, or its molecules occupy positions of advantage, so that the energy expended can be at any time recovered by the return of the body to its original position, or by the return of its molecules to their original positions.

85. Unit of Work and Energy. — Inasmuch as a body's capacity to do work is dependent wholly upon the work which has been done upon it, it is evident that both work and energy may be measured by the same unit. The unit adopted is *the work done or energy imparted in raising one pound through a vertical hight of one foot.* It is called a *foot-pound*. (The French unit is the work done or energy imparted in raising 1^k to a vertical hight of

1^m, and is called a kilogrammeter.) Since the work done in raising 1 pound 1 foot high is 1 foot-pound, the work of raising it 10 feet high is 10 foot-pounds, which is the same as the work done in raising 10 pounds 1 foot high; and the same, again, as raising 2 pounds 5 feet high.

In this unit, and by means of formulas (1) and (2), page 99, we are able to estimate any species of work, and thereby compare work of any kind with that of any other kind. For instance, let us compare the work done by a man in sawing through a stick of wood, whose saw must move 100 feet (S) against an average resistance (R) of 20 pounds, with that done by a bullet in penetrating a plank to the depth of 2 inches ($\frac{1}{6}$ foot) against an average resistance of 500 pounds. Moving a saw 100 feet against a resistance of 20 pounds is equivalent to raising 20 pounds 100 feet high, or doing (RS = 20 × 100 =) 2,000 foot-pounds of work (W); a bullet moving $\frac{1}{6}$ foot against 500 pounds' resistance does the same amount of work as is required to raise 500 pounds $\frac{1}{6}$ foot high, or (500 × $\frac{1}{6}$ =) 83.3 + foot-pounds of work. Hence (2,000 ÷ 83.3 =) about 24 times as much work is done by the sawyer as by the bullet.

<small>Let us estimate the energy stored in a bow, by an archer whose hand in pulling on the string, while bending the bow moves 6 inches ($\frac{1}{2}$ foot) against an average resistance of 20 pounds. Here RS = 20 × $\frac{1}{2}$ = 10 foot-pounds of work done upon the bow, or 10 foot-pounds of energy stored in the bow.</small>

86. Distinction between Force and Energy. — Force may be measured by an instrument called a dynamometer. Energy which is the product of force into space *cannot* be measured directly by *any* instrument. Force can be increased indefinitely by means of machines, as a lever, hydrostatic press, etc.; *energy cannot be increased by any instrument or means whatsoever.*

87. Formula for Calculating Kinetic Energy.

— The kinetic energy of a moving body is calculated by means of the following formula: —

$$\frac{WV^2}{2g} = \text{energy},$$

in which W represents the weight of the body, V its velocity, and g the acceleration (in this latitude $32\frac{1}{6}$ feet, or 9.8^m per second) due to gravity. [For the derivation of this formula, see the Author's Elements of Physics, pages 123 and 124.] For example, the energy of a cannon-ball weighing 50 pounds and moving with a velocity of 1,000 feet per second $= \frac{WV^2}{2g} = \frac{50 \times 1000^2}{2 \times 32\frac{1}{6}} =$ (about) 779,301 foot-pounds.

Certain deductions from this formula should be strongly impressed upon the mind; viz., (1) *with the same velocity the kinetic energy of a body varies as its weight;* (2) *with the same weight its kinetic energy varies as the square of its velocity.* Doubling the velocity multiplies the energy fourfold; trebling the velocity multiplies it ninefold. A bullet moving with a velocity of 600 feet per second will penetrate, not twice, but four times, as far into a plank as one having a velocity of 300 feet per second.

A railway train having a velocity of 20 miles an hour will, if the steam is shut off, continue to run four times as far as it would if its velocity were 10 miles an hour. The reason is apparent why light substances, even so light as air, exhibit great energy when their velocity is great.

88. Wasted Work.

— As a stone is raised higher and higher, the work *accumulates* in the form of potential energy. As a body free to move (*i.e.* meeting with no resistance) acquires, under the influence of a constant force, uniformly accelerated motion, so does work, in the form of kinetic energy, accumulate. But accumulated work or energy does not always vary as the work performed. In practice, more or less of the work done, especially that done in overcoming friction, resistance of fluids, etc., is *wasted*. The work done by the sawyer and bullet, page 102, so far as imparting energy to the bodies on which they do work, is all lost. Of the vast amount of work done in propelling a steamer across the ocean none accumulates; all is wasted, distributed along the watery path, and cannot be recovered or made available for doing more work. Evidently *the accumulated work or available energy that a body possesses is the work done upon the body less the wasted work*. We may then calculate in foot-pounds (or kilogrammeters) according to formulas (1) or (2), page 99, the work performed on a body, and from this deduct the number of foot-pounds wasted; the remainder is the number of foot-pounds of available energy that is imparted to the body.

89. Power of an Agent to do Work, or Rate at which an Agent can do Work.—In estimating the total amount of work done, the time consumed is not taken into consideration. The work done by a hod-carrier, in carrying 1,000 bricks to the top of a building, is the same whether he does it in a day or a week. But in estimating the power of any agent to do work, as of a man, a horse, or a steam-engine, in other words, *the rate* at which it is capable of doing work, it is evident that time is an important element. The work done by a horse, in raising a barrel of flour 20 feet high, is about 4,000 foot-pounds; but even a mouse could do the same amount of work in time.

The unit in which power or rate of doing work is estimated is called (inappropriately) a *horse-power*. *A horsepower represents the power to perform* 33,000 *foot-pounds of work in a minute* (or 550 foot-pounds in a second).

EXERCISES.

1. Can a person lift himself, or put himself in motion, without exerting force upon some other body?

2. Can a body do work upon itself? Can a body generate energy in itself, *i.e.* increase its own energy?

3. (*a*) Suppose that an average force of 25 pounds is exerted through a space of 10 inches in bending a bow; what amount of energy will it give the bow? (*b*) What kind of energy will the bow, when bended, possess?

4. (*a*) What amount of kinetic energy does a body weighing 20 pounds, and moving with a velocity of 300 feet per second, possess? (*b*) What amount of work can the body do?

5. (*a*) What amount of work is required to raise 50 tons of coal from a mine 200 feet deep? (*b*) An engine of how many horse-power would be required to do it in two hours?

6. How many fold is the kinetic energy of a body increased by doubling its velocity?

7. Twelve hundred foot-pounds of energy will raise 100 pounds how high, if none is wasted?

8. A force of 500 pounds acts upon a body through a space of 20 feet. One-fourth of the work is wasted in consequence of resistances. How much available energy is imparted to the body?

9. How much energy is stored in a body weighing 1,000 pounds, at a hight of 200 feet above the earth?

10. How much work can a 2 horse-power engine do in an hour?

11. A horse draws a carriage on a level road at the uniform rate of 5 miles an hour. (*a*) Does work accumulate? (*b*) What kind of energy does the carriage possess? (*c*) Suppose that the carriage were drawn up a hill; would energy accumulate? (*d*) What kind of energy would it possess when at rest on the top of the hill? (*e*) How would you calculate the quantity of energy it possesses when at rest on top of the hill? (*f*) Suppose that the carriage is in motion on top of the hill; what two formulas would you employ in calculating the total energy which it possesses?

Section II.

THE ABSOLUTE OR C.G.S. SYSTEM OF MEASUREMENTS.

90. Fundamental Units. — For many scientific purposes, especially in establishing a complete set of electrical units, a different system for measuring physical quantities than that in common use and called the *gravitation system*, is indispensable. In the new system, all physical quantities may be expressed in terms of *three* units, which are called *fundamental units*. All others are deduced from these by definition, and are called *derived units*. The fundamental units and their symbols are as follows: —

Unit of length, L: the *centimeter*, or the hundredth part of a meter.

Unit of mass, M: the *gram*, or the mass of one cubic centimeter of distilled water at 4° C.

Unit of time, T: a *second*.

The system of units based on these three fundamental units is called the Absolute System, or the Centimeter-Gram-Second System, or, by abbreviation, C.G.S. System.

91. Derived Units. — There are a great number of derived units. We give a few of those in most common use.

Unit of velocity, V: one centimeter per second; in uniform motion,

$$V = \frac{S}{T}.$$

Unit of acceleration, A: an increase of velocity of one centimeter per second.

Unit of force, F: *a dyne*; it is that force which, acting for a second, will give to a mass of one gram an acceleration of one centimeter per second, *i.e.* one unit of acceleration. It is the $\frac{1}{g}$ part of the weight of the unit of mass.

$$F = MA = \frac{ML}{T^2}, \text{ or } MLT^{-2}.$$

Unit of work, W; or of energy, E: an *erg*; it is the work done or energy imparted by a force of one dyne working through a length or distance of one centimeter.

$$W \text{ or } E = FS = MAS = \frac{ML^2}{T^2}, \text{ or } ML^2T^{-2}.$$

92. Relation of the Dyne to the Gram or Gravitation Unit of Weight. — When a body falls in a vacuum, gravity imparts to it an acceleration of g (in the latitude of the Northern States, 980) centimeters per second. The force of gravity, therefore, acting on a unit of mass is, according to definition, g (980) dynes. The weight of a mass of one gram is in the gravitation system one gram. Hence the gram (gravitation unit of weight) must be equal to g dynes, or, in the Northern United States, to 980 dynes. The weight of a mass of one gram varies with the latitude and hight above the earth's surface, while *the mass of a gram and the dyne are constant quantities*; their value does not change with place.

93. Another Formula for Computing Kinetic Energy.

— It is evident that the weight of a body is dependent upon its mass and the force of gravity; in other words, (W = Mg) the weight of a body is measured by the product of the acceleration which the force of gravity produces into its units of mass. Hence the mass of a body is numerically $\frac{1}{g}$ its weight. Substituting the value of W given above in the formula (p. 103), $E = \frac{WV^2}{2g}$, we have $E = \frac{MV^2}{2}$. When the latter formula is used, it is evident that the *mass* of the moving body must be found by dividing its weight in grams by 980, or its weight in pounds by 32.1 +.

The absolute system is used in all refined physical measurements, but the gravitation system is more convenient and is universally used in the ordinary affairs of life.

QUESTIONS.

[Designed for only those who may take up the absolute system.]

1. (*a*) Name some unit of force which is based upon the weight of some definite mass. (*b*) Name some unit of force which is based upon the amount of acceleration which a force can impart to a body of a given mass in a given time. (*c*) Have both of these units *absolute* (unchangeable) values? (*d*) What names do you employ in distinguishing these two classes of units?

2. (*a*) What are the fundamental units of the absolute system? (*b*) Why are they called fundamental units?

3. A force of 20^g is equivalent to how many dynes?

4. (*a*) A force of 20 dynes would perform how many ergs of work in acting through a distance of 10^{cm}? (*b*) How many ergs of work would a force of 20 grams perform in acting through the same distance? (*c*) How many kilogrammeters of work would a force of 20 grams perform in acting through the same distance?

5. What is the weight of a mass of 1^g in dynes?

Section III.

MACHINES.

94. Uses of Machines.

Experiment 74.— Suspend, as in Figure 95, a fixed pulley, A, and a movable pulley, B. The scale-pan C counterbalances the pulley B, so that there will be equilibrium. Suspend from B two balls, LL, of equal weight, and suspend on the side where the pan is, a single ball,

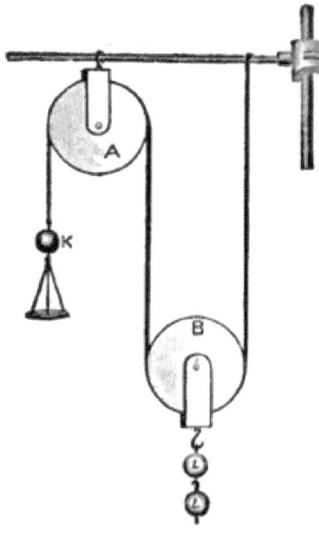

Fig. 95.

K, equal to one of the former. The single ball supports the two balls; *i.e.* by the use of the machine, a force of 1 is enabled to balance a force of 2. So far no work is done. (Why?) Place a very small weight in the pan, and the balls LL begin to rise, and work is done.

As the weight K plus a very small weight causes the motion, we shall regard this as the force (F); and as the weights LL are the bodies moved (the pulleys and pan being parts of the machine may be disregarded), they may be regarded as the resistance (R) overcome, or the body on which work is done. Measure the respective distances through which F acts and R moves during the same time. R moves only one-half as great a distance as that through which F acts; *i.e.* if R rises 2 feet, F must act through 4 feet. Suppose that R is 2 pounds, then F is 1 + pounds. Now 2 (pounds) × 2 feet = 4 foot-pounds of work done on R. Again, 1 + (pounds) × 4 feet = a little more than 4 foot-pounds of work (or energy) expended.

It thus seems that, although a machine will enable a small force to balance a large force, when work is performed, the work applied to the machine is greater, rather than less, than the work which the machine transmits to the resistance. The work applied is greater than the work transmitted by the amount of work wasted in consequence of friction and other extra resistances. So that *by the employment of a machine nothing is gained in work which the force is required to do, but always something lost.*

What, then, is the advantage gained in using this machine? Suppose that R is 400 pounds, and that the utmost force that a man can exert is a little more than 200 pounds. Then, without the machine, the services of two men would be required to move the resistance; whereas one man can move it with a machine, only that he will be obliged to move twice as far as the resistance moves, a matter of little consequence in comparison with the advantage of being able to do the work alone. The advantage gained in this instance seems to be one of *convenience.* Men, however, are accustomed to speak of it as "*a gain of force,*" (or more commonly and inaccurately, "*of power*"), inasmuch as a small force overcomes a large resistance.

Experiment 75. — If instead of applying the small additional weight to the pan, it be suspended from one of the balls LL, the weight of these balls, together with the additional weight, becomes the cause of motion, and K is the resistance. In this case there is a loss of force, because the force employed is more than twice as great as the force overcome. Measure the distances traversed respectively by F and R in the same time. R moves twice as far as F, and of course with twice the velocity. There is *a gain of velocity* at the expense of *force*.

It thus appears that, if it should be desirable to move a resistance with greater velocity than it is possible or convenient for the force to act, it may be accomplished through the mediation of a machine, by applying to it a force proportionately greater than the resistance. This apparatus is one of many *contrivances called machines, through the mediation of which force can be applied to resistance more advantageously than when it is applied directly to the resistance.* Some of the many advantages derived from the use of machines are: —

(1) *They may enable us to exchange intensity of force for velocity, or velocity for intensity of force.* A gain of intensity of force or a gain of velocity is called a *mechanical advantage.*

(2) *They may enable us to employ a force in a direction that is more convenient than the direction in which the resistance is to be moved.*

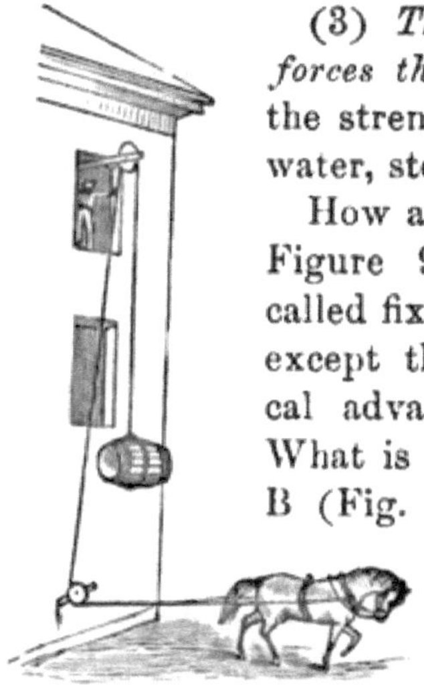

(3) *They may enable us to employ other forces than our own in doing work;* e.g. the strength of animals; the forces of wind, water, steam, etc.

How are the last two uses illustrated in Figure 96? The pulleys employed are called fixed pulleys, *i.e.* they have no motion except that of rotation. Is any mechanical advantage gained by fixed pulleys? What is the use of a fixed pulley? Pulley B (Fig. 94) is a movable pulley. What kind of advantage is gained by means of a movable pulley?

Fig. 96.

95. General Law of Machines. — From the experiments and discussion above we derive the following formula for machines: —

$$FS = RS' + w,$$

in which F represents the force applied, and S the distance through which F acts; R represents the resistance overcome, and S′ the distance through which its point of application is moved; *w* represents the wasted work. A machine in which there is no wasted work is a *perfect machine*. Such a machine is purely ideal, as none exists. If in our calculations we regard a machine as perfect (though subsequently suitable allowance must be made for the wasted work), then our formula becomes

$$FS = RS'.$$

Whence $R : F :: S : S'$; i.e. *the force and resistance vary inversely as the distances which their respective points of application move*. In other words, the ratio of the resistance to the force is the reciprocal of the ratio of the distances which these points move.

$$R : F = 4, \text{ then } S' : S = \tfrac{1}{4}.$$

This law applies to every machine of whatever description; hence it is called the *General* or *Universal Law of Machines*. When R is greater than F, there is a gain of force, and $\frac{R}{F} =$ *the ratio of gain of force*. When S′ is greater than S, there is a gain of velocity, and $\frac{S'}{S} =$ *the ratio of gain of velocity*.

Experiment 76. — Support a lever, as in Figure 97, so that there shall be unequal arms. Move *w* until the lever is balanced in a horizontal position. Suspend (say) seven balls from the short arm (say) one space from the fulcrum. Then from the other arm suspend a single ball from such a place (in this case seven equal spaces from the fulcrum) that it will balance the seven balls. There is now equilibrium between the two forces. Suppose

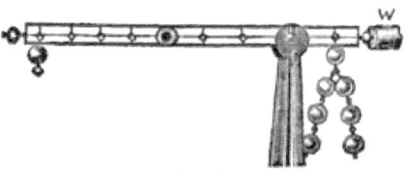

Fig. 97.

the smaller force to be increased a little and to produce motion; what mechanical advantage (*i.e.* intensity of force or velocity) would be gained by the use of the machine? What is the ratio of gain neglecting the small additional force? How does this ratio compare with the ratio between the length of the two arms? For convenience we call the distance of the point of application of the force from the fulcrum the *force-arm*, and the distance of the resistance from the fulcrum the *resistance-arm*.

Suppose the small additional force is applied to the short arm; what mechanical advantage would be gained? What would be the ratio of gain?

While the general law of machines is always applicable, a *special law*, one in which the relation between the ratio of gain and the ratio between certain dimensions of the machine is stated, is often more convenient in practice. For example, in our experiment with the lever we discover that R : F :: force-arm : resistance-arm, *i.e. the force and resistance vary inversely as the lengths of their respective arms.* Compare this special law with the general law.

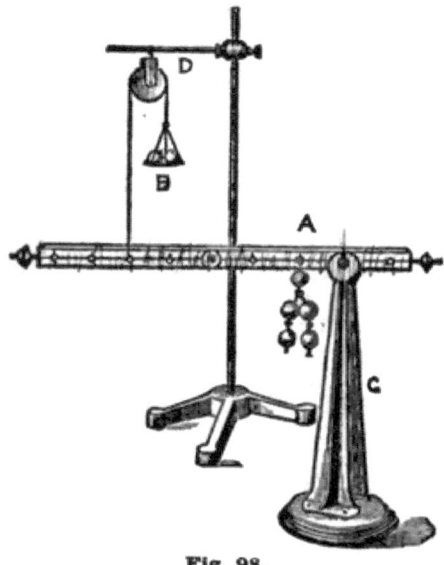

Fig. 98.

Place the fulcrum at other points in the lever, and thereby vary the length of the arms, and verify by numerous experiments the special law of levers.

Experiment 77. — By means of a pulley, D, so arrange (Fig. 98) that both F and R may be on the same side of the fulcrum. First,

MACHINES. 113

place in the pan weights sufficient to produce equilibrium in the machine (for example, in this case, one ball). Then suspend weights at some point, as A, and place other weights in the pan to counterbalance these. Verify the law of levers. If A is the resistance, what mechanical advantage is gained? What is the ratio of gain? If B is the resistance, what mechanical advantage will be gained?

Experiment 78.—Obtain a toy carriage, place it on an inclined plane, pass the cord over a pulley, B (Fig. 99), so adjusted that the cord between the carriage and pulley shall be parallel with the plane. Suspend a small bucket, P, and place sand in it to balance the carriage.

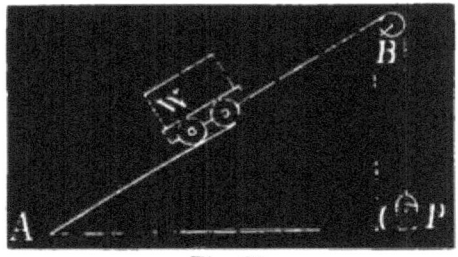

Fig. 99.

Place in the carriage a weight W, and place weights in the bucket to balance W. The weights placed in the bucket represent the force

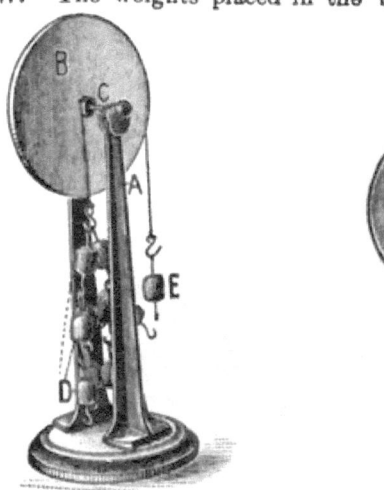

Fig. 100.

Fig. 101.

applied; then what advantage is gained in the use of an inclined plane as a machine? W, in traversing the inclined plane AB, only rises through the vertical hight CB, while P must move through a distance equal to AB. Measure the distances AB and CB. How does the ratio between these distances compare with the ratio of gain? Construct a special law of the inclined plane.

Experiment 79. — Place a "wheel and axle" (Fig. 100) on the support A. Wind a cord around the wheel B, and another in the reverse direction around the axle C. Suspend a weight, D, from the axle, and another, E, from the wheel, to balance it. If E be the force applied, what advantage is gained? What, if D is the force applied? What is the ratio of advantage in either case? How does the ratio of advantage compare with the ratio between the radius of the wheel AC (Fig. 101) and the radius of the axle BC? Construct a special law of the wheel and axle.

Fig. 102.

EXERCISES.

1. (a) When is a machine said to gain intensity of force? (b) When is it said to gain velocity?

2. (*a*) Can any machine do work? (*b*) Can we by the use of any machine accomplish more work than the work performed upon the machine? What is the proof?

3. How is intensity of force gained by the use of a machine?

4. What machine is used only to change the direction of motion?

Fig. 103.

5. (*a*) What is a mechanical advantage? (*b*) Give a rule by which the mechanical advantage that may be gained by any machine may be calculated.

Fig. 104.

6. Figure 102 represents a pile-driver. (*a*) How can the energy or the work which the weight W can do when it is raised a given distance be computed? (*b*) What benefit is derived from the use of the machine in raising the weight? (*c*) Suggest some simple attachment to the machine which would enable one man to raise the weight. (*d*) Suggest some attachment by means of which a horse could be made to do the work. (*e*) What difference will it make whether the weight is raised 5 feet or 10 feet? (*f*) Illustrate, by means of this machine, what you understand by force and energy. (*g*) Which, while the weight rises, is constantly accumulating, and which remains nearly constant? (*h*) Which can be meas-

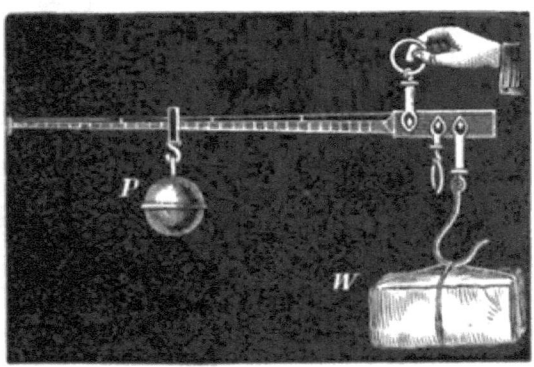

Fig. 105.

ured with an instrument, and what is the name of the instrument?

7. (*a*) What advantage is gained by a lever when its force-arm is longer than its resistance-arm? (*b*) What, when its resistance-arm is longer?

8. (*a*) What advantage is gained by a nut-cracker (Fig. 103)? (*b*) What is the ratio of gain?.

9. (a) What advantage is gained by cutting far back on the blades of shears near the fulcrum? Why? (b) Should shears for cutting metals be made with short handles and long blades, or the reverse? (c) What is the advantage of long blades?

Fig. 106.

10. Is work done when the moment of the force applied to a lever is equal to the moment of the resistance? Why?

11. (a) If P (Fig. 105), weighing 1 pound, is suspended 15 spaces from the fulcrum of the steelyard, what weight (W), suspended 3 similar spaces the other side of the fulcrum, will balance it? (b) Where would you place the one-pound weight in order to weigh out 6 pounds of tea?

12. (a) If the circumference of the axle (Fig. 106) is 15 inches, and the force applied to the crank acts through 15 feet during each revolution, what force will be necessary to raise the bucket of coal weighing (say) 36 pounds? (b) Through how many feet must the force act to raise the bucket from a cavity 48 feet deep?

13. The arm is raised by the contraction (shortening by muscular force) of the muscle A (Fig. 107), which is attached at one extremity to the shoulder and at the other extremity B to the fore-arm, near the

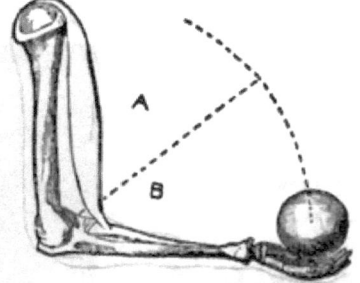

Fig. 107.

elbow. (a) When the arm is used, as represented in the figure, to raise a weight, what kind of a machine is it? (b) What mechanical advantage is gained by it? (c) How can the ratio of gain be computed? (d) For which purpose is the arm adapted, to gain intensity of force or velocity?

The lengths of the two arms of a balance, such as is used in finding specific gravity (Fig. 60, page 61), should be exactly equal. The arms may be of unequal length, and yet the beam may be in equilibrium (*i.e.* take a horizontal position when no weights are applied), in consequence of having more matter in the shorter arm, as in Figure 97, page 111. Such a balance is called a *false balance.*

14. (*a*) How would you test a balance to ascertain whether it is true or false? (*b*) If you were buying diamonds, and the seller should sell them to you by weight as obtained by placing them on the shorter arm of a false balance, would you be the loser or gainer?

The true weight of a body may be found with a false balance by a process called *double weighing.* The article to be weighed is placed in one pan, and a counterpoise of sand in the other pan. The article is then removed, and known weights placed in the pan until equilibrium is again produced. These weights represent the correct weight of the article. In this way the balances used in the school laboratory should be tested by the pupil.

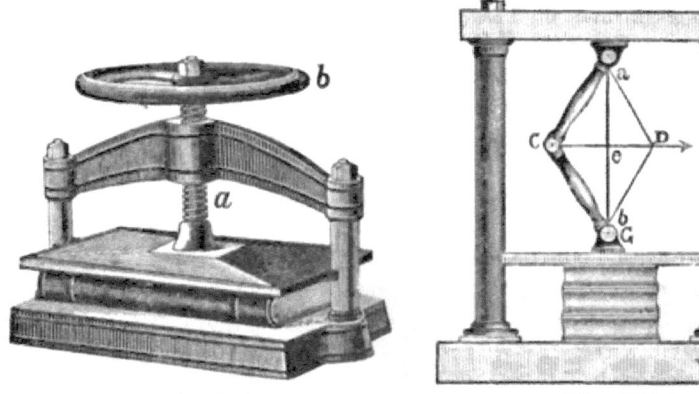

Fig. 108.   Fig. 109.

15. During one revolution a screw advances a distance equal to the distance between two threads, measured in the direction of the axis of the screw. Suppose the screw in the letter-press (Fig. 108) to advance $\frac{1}{4}$ inch at each revolution, and a force of 25 pounds to be applied to the circumference of the wheel B, whose diameter is 14 inches. What pressure would be exerted on articles placed beneath the screw? (The circumference of a circle is 3.1416 times its diameter.)

16. The toggle-joint (Fig. 109) is a machine employed where great pressure has to be exerted through a small space, as in punching and

shearing iron, and in printing-presses, in pressing the types forcibly against the paper. An illustration may be found in the joints used to raise carriage-tops. Force applied to the joint C will cause the two links AC and BC to be straightened, or carried forward to e. If point C moves 5 inches while G moves ½ inch, then what pressure will a force of 50 pounds applied at C exert on the book below?

Fig. 110.

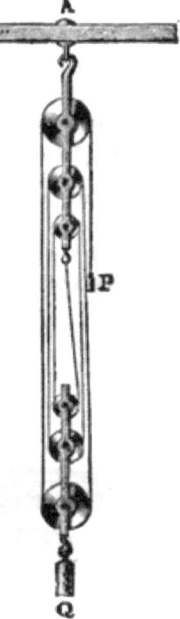

Fig. 111.

17. Show that the hydrostatic press (page 50) conforms in its operation to the general law of machines.

18. A wedge may be regarded as two inclined planes placed base to base, as dc (Fig. 110). (a) What mechanical advantage is gained by it? (b) Suppose that the thickness ab is 4 inches, and the length dc is 8 inches, and that the average pressure exerted upon it by the blow of a sledge is 100 pounds; what will be the average pressure exerted by the wedge tending to separate the fibers of wood?

A compound machine is one consisting of two or more machines combined in one; e.g. compound pulleys (Fig. 111) and compound wheels and axles (Fig. 112). The mechanical advantage that may be gained by a compound machine may be calculated by multiplying continuously together the ratios of the several machines.

Fig. 112.

19. (a) How great is the advantage gained by one movable pulley? (b) How great is the advantage gained by the compound pulley (Fig. 111) consisting of three movable pulleys?

20. Suppose that the radii of the wheels a, d, and f (Fig. 112) are, respectively, 20 inches, 16 inches, and 24 inches, and the radii of their axles are, respectively, 2 inches, 4 inches, and 6 inches; how great advantage may be gained by the compound machine?

Fig. 113.

Fig. 114.

21. How would you calculate the mechanical advantage gained by a machine like that of Figure 113? (On the axle A is an endless screw, by means of which motion is communicated from the axle to the wheel W.)

22. (a) What kind of a machine is a claw-hammer (Fig. 114)? (b) What mechanical advantage is gained by it?

23. In its technical meaning, a "perpetual motion machine" is not a machine that will run indefinitely, but a machine which *can do work without the expenditure of energy*. Is such a machine possible?

24. A plank 12 feet long and weighing 24 pounds is supported by two props, one 3 feet from one end, and the other 1 foot from the other end. What is the pressure on each prop?

25. With a movable pulley what force will support a weight of 100 pounds?

26. The gradient of a certain road on a hillside is one foot in ten feet. What force must a horse exert on a carriage which weighs together with its load one ton, to prevent its descent?

27. What must be the diameter of a wheel in order that a force of 20 pounds applied at its circumference may be in equilibrium with a resistance of 600 pounds applied to its axle, which is 3 inches in diameter?

120   WORK AND ENERGY.

28. Draw a straight line to represent a lever; locate the fulcrum, and locate the points of application of the force and resistance unequally distant from the fulcrum. Draw lines from the points of application of the force and resistance so that they will make some angle with each other (*i.e.* not parallel with each other) to represent the directions in which the two forces respectively act. Ascertain the ratio between the two forces when their moments are equal, *i.e.* when the two forces are in equilibrium.

CHAPTER V.

MOLECULAR ENERGY.—HEAT.

Section I.

WHAT HEAT IS.—SOME SOURCES OF HEAT.

96. Theory of Heat.—A body loses motion in communicating it. The hammer descends and strikes the anvil; its motion ceases, but the anvil is not sensibly moved; the only observable effect produced is heat. Instead of a motion of the hammer and anvil, there is now, according to the modern view, an increased *vibratory* motion of the *molecules* that compose the hammer and anvil, —*simply a change from molar to molecular motion.* Of course, this latter motion is invisible. According to this view, *heat is but a name for the energy of vibration of the molecules of a body.* A body is heated by having the motion of its molecules quickened, and cooled by parting with some of its molecular motion. *One body is hotter than another when the average kinetic energy of each molecule in it is greater than in the other.*

As late as the beginning of the present century heat was generally regarded as "a sensation which the presence of fire" (an "igneous *fluid*," "matter of heat," called sometimes "caloric") "occasions in animate and inanimate bodies." A text-book of that period makes this significant statement: "There is fire in the wood, and there is air in the field, though we do not perceive either while at rest. Rubbing two pieces of wood does not create fire any more than the blowing of the wind creates air. *Motion renders both perceptible.*" The former and the more modern views are in harmony in attributing the immediate cause of the sensation to *motion*. According to the former view, the sensation is produced by *putting an imaginary fluid in motion*; according to the modern view it is produced by *quickening the motion of the molecules of a body*.

97. Artificial Sources of Heat. — As heat is energy, so *all heat must originate in some form of energy*, i.e. *by the transformation of some other form of energy into heat.*

Experiment 80. — Place a ten-penny nail on a stone or a flat piece of iron and hammer it briskly for a few minutes. It soon becomes too hot to be handled with comfort. Rub a desk with your fist; your coat-sleeve with a metallic button; both the rubbers and the things rubbed become heated.

(1) Heat is generated at the expense of molar motion, *i.e. molar motion checked becomes molecular motion, or heat.*

Experiment 81. — Take a glass test-tube half full of cold water and pour into it one-fourth its volume of strong sulphuric acid. The liquid almost instantly becomes so hot that the tube cannot be held in the hand.

When water is poured upon quicklime, heat is rapidly developed. The invisible oxygen of the air combines with the constituents of the various fuels, such as wood, coal, oils, and illuminating-gas, and gives rise to what we call *burning*, or *combustion*, by which a large amount of heat is generated. In all such cases the heat is generated by the combination or clashing together of molecules of substances that have an affinity (*i.e.* an attraction) for one another. Before union the energy of the molecules is of the same kind as that of a stone on a shelf. When the shelf is withdrawn, gravity converts the potential energy of the stone into kinetic energy; so affinity converts the potential energy of the molecules into kinetic energy of vibration; *i.e.* into heat.

(2) *Molecular (or atomic) potential energy is transformed in the act of chemical combination into heat.*

98. The Sun as a Source of Heat and Energy. — The sun is the source of very nearly all the energy employed by man in doing work. Our coal-beds, the results of the deposit of vegetable matter, are vast storehouses of the sun's energy, rendered potential during the growth of the plants many ages ago. The animal finds its food in the plant, appropriates the energy stored in the plant, and converts it into energy of motion in the form of *animal heat* and *muscular motion*. Every rain-drop that rolls its way to the sea, contributing its mite to the immense *water-power* of the earth, derives its energy from the sun.

QUESTIONS.

1. On every hand we see what appears to be at least an almost universal tendency to destruction of motion. Is the destruction usually an annihilation of motion?
2. What name is usually given to molecular energy?
3. How does it appear that heat is energy?
4. What do you mean when you say that one body is hotter than another?
5. How must all heat originate?
6. State all the sources of heat with which you are now acquainted.
7. (*a*) Give an illustration of mechanical or visible motion transformed into molecular motion. (*b*) Give an illustration of molecular motion transformed into mechanical motion.
8. What kind of energy does coal and other fuel possess?
9. A lump of coal is raised and placed upon a shelf. (*a*) How can the potential energy of the lump be transformed into kinetic energy? (*b*) Will the kinetic energy resulting from the transformation be mechanical or molecular? (*c*) When the lump strikes the earth, what transformation of energy occurs?
10. Every lump of coal possesses molecular potential energy. (*a*) How can its energy be transformed into kinetic energy? (*b*) What

two varieties of potential energy does a lump of coal on the shelf possess?

11. (*a*) How do bodies acquire energy? (*b*) From what source did coal obtain its molecular potential energy? (*c*) What does the entire value of coal consist in?

12. How does animal energy originate?

Section II.

TEMPERATURE. — METHODS OF EQUALIZATION.

99. Temperature Defined. — If body A is brought in contact with body B, and A tends to impart heat to B, then A is said to have a higher *temperature* than B. *Temperature is the state of a body with reference to its tendency to communicate heat to, or receive heat from, other bodies.* The direction of the flow of heat determines which of two bodies has the higher temperature. If the temperature of neither body rises at the expense of the other, then both have the same temperature.

100. Temperature distinguished from Quantity of Heat. — The term *temperature* does not signify *quantity of heat*. If we dip from a gallon of boiling water a cupful, the cup of water is just as hot, *i.e.* has the same temperature, as the larger quantity, although of course there is a great difference in the quantities of heat the two bodies of water contain. *Temperature depends upon the average kinetic energy of the individual molecule, while quantity of heat depends upon the average kinetic energy of the individual molecule multiplied by the number of molecules.*

There is always a tendency to *equalization of temperature;* that is, heat has a tendency to pass from a warmer body to a colder, or from a warmer to a colder part of the same body, until there is an equality of temperature.

101. Conduction.

Experiment 82. — Place one end of a wire about 10 inches long in a lamp-flame, and hold the other end in the hand. Heat gradually travels from the end in the flame toward the hand. Apply your fingers successively at different points nearer and nearer the flame; you find that the nearer you approach the flame, the hotter the wire is.

The flow of heat through an unequally heated body, from places of higher to places of lower temperature, is called *conduction;* the body through which it travels is called a *conductor.* The molecules of the wire in the flame have their motion quickened; they strike their neighbors and quicken their motion; the latter in turn quicken the motion of the next; and so on, until some of the motion is finally communicated to the hand, and creates in it the sensation of heat.

Experiment 83. — Figure 115 represents a board on which are fastened, by means of staples, four wires: (1) iron, (2) copper, (3) brass, and (4) German silver. Place a lamp-flame where the wires meet. In about a minute run your fingers along the wires from the remote ends toward the flame, and see how near you can approach the flame on each without suffering from the heat. Make a list of these metals, arranging them in the order of their conductibility.

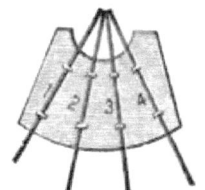

Fig. 115.

You learn that some substances conduct heat much more rapidly than others. The former are called *good conductors,* the latter *poor conductors.* Metals are the best conductors, though they differ widely among themselves.

Experiment 84. — Fill a test-tube full of water, and hold it somewhat inclined (Fig. 116), so that a flame may heat the part of the tube near the surface of the water. Do not allow the flame to touch the part of the tube that does not contain water. The water may be made to boil near its surface for several minutes before any change of the temperature at the bottom will be perceived.

Fig. 116.

Liquids, as a class, are poorer conductors than solids. Gases are much poorer conductors than liquids. It is difficult to discover that pure, dry air possesses any conducting power. The poor conducting power of our clothing is due partly to the poor conducting power of the fibers of the cloth, but chiefly to the air which is confined by it.

Loose garments, and garments of loosely woven cloth, inasmuch as they hold a large amount of *confined air*, furnish a good protection from heat and cold. Bodies are surrounded with bad conductors, to *retain* heat when their temperature is above that of surrounding objects, and to *exclude* it when their temperature is below that of surrounding objects. In the same manner double windows and doors protect from cold.

102. Convection in Gases.

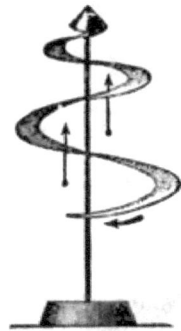

Fig. 117.

Experiment 85. — Hold your hand a little way from a flame, beneath, on the side of, and above the flame. At which place is the heat most intense?

Experiment 86. — Draw on thin glazed paper an unfolding line, so that the windings shall be about ⅜ inch apart. Cut along the line; give the central portion a conical form; place the cone on a pointed end of a vertical wire, and allow the remainder of the paper to fall spirally around the wire as in Figure 117. Place the spiral over a flame or hot stove. A continuous current of air, a miniature wind, moving upward from the flame or stove causes the spiral to rotate. This current tends only upward. The air having become heated by contact with the surfaces of the flame or stove conveys, in its ascent, heat to objects above. Heat is thus diffused by a process called *convection* (conveying).

Experiment 87. — Cover a candle-flame with a glass chimney (Fig. 118), blocking the latter up a little way so that there may be a circulation of air beneath. Hold the spiral over the chimney; the rotation is much quicker than before. Hold smoking touch-paper near the bottom of the chimney; the smoke seems to be *drawn* with great rapidity into the chimney at the bottom; in other words, the office of the chimney is to create what is called a *draft* of air. Notice whether the combustion takes place any more rapidly with than without the chimney.

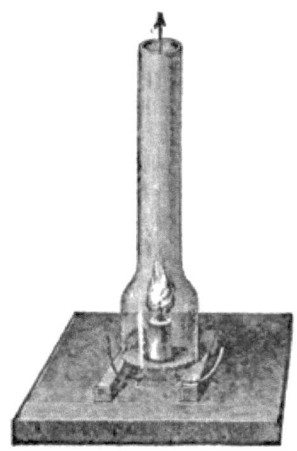

Fig. 118.

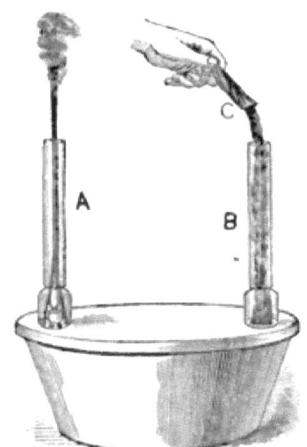

Fig. 119.

Experiment 88. — Place a candle within a circle of holes cut in the cover of a vessel, and cover it with a chimney, A (Fig. 119). Over an orifice in the cover place another chimney, B. Hold a roll of smoking touch-paper over B. The smoke descends this chimney, passes through the vessel and out at A. This illustrates the method often adopted to produce a ventilating draft through mines. Let the interior of a tin vessel represent a mine deep in the earth, and the chimneys two shafts sunk to opposite extremities of the mine. A fire kept burning at the bottom of one shaft will cause a current of air

128   MOLECULAR ENERGY. — HEAT.

to sweep down the other shaft, and through the mine, and thus keep up a circulation of pure air through the mine.

The cause of the ascending currents is evident. Air, on becoming heated, expands rapidly and becomes much rarer than the surrounding colder air; hence it rises much like a cork in water, while cold air pours in laterally to take its place. In this manner winds are created.

The so-called *trade-winds* originate in the torrid or heated zone of the earth. The air over the heated surface of the earth rises, and the colder air from the polar regions flows in on both sides, giving rise to a constant southward wind in the northern hemisphere, and northward wind in the southern hemisphere.

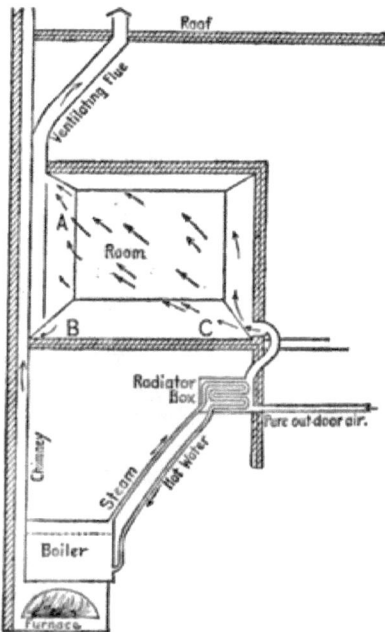

Fig. 120.

Chemistry teaches us the vital importance of thorough ventilation. Figure 120 represents a scheme for heating a room by steam, and ventilating it by convection. Steam is conveyed by a pipe from the boiler to a radiator box just beneath the floor of the room. The air in the box becomes heated by contact with and radiation from the coil of pipe in the box, and rises through a passage opening by means of a register into the room near the floor at C, a supply of pure air being kept up by means of a tubular passage opening into the box from the outside of the building. Thus the room is furnished with *pure warm* air, which, mingling with the impurities arising from the respiration of its occupants, serves to dilute them, and render them less injurious. At the same time, the warm and partially vitiated air of the room passes through the open ventilator, A, into the ventilating-flue, and escapes, so that in a moderate length of time a nearly complete change of air is effected. It is evident that on the coldest days of winter the convection is most rapid; indeed, it may be so rapid that the air cannot be heated sufficiently to render the room near the floor comfortable. At such times the ventilator A may be closed, while the ventilator B is always open. The heated air rises to top of the room and, not being able to escape, crowds the colder air beneath out at the ventilator B. No system of ventilation dependent wholly on convection is adequate to ventilate properly crowded halls; air is too viscous and sluggish in its movements. In such cases ventilation should be assisted by some mechanical means, such as a blower or fan, worked by steam or water power.

103. Convection in Liquids.

Experiment 89. — Fill a small (6 ounce), thin glass flask with boiling hot water, color it with a teaspoonful of ink, stopper the flask, and lower it deep in a tub, pail, or other large vessel filled with cold water. Withdraw the stopper, and the hot, rarer, colored water will rise from the flask, and the cold water will descend into the flask. The two currents passing in and out of the neck of the flask are easily distinguished. The colored liquid marks distinctly the path of the heated convection currents through the colored liquid and makes clear the method by which heat, when applied at the bottom of a body of liquid, becomes rapidly diffused through the entire mass notwithstanding that liquids are poor conductors.

Experiment 90. — Fill again the flask with hot colored water, stopper, invert, and introduce the mouth of the flask just beneath the surface of a fresh pail of cold water. Withdraw the stopper with as little agitation of the water as possible. What happens? Explain.

104. Radiation. — In some way the sun is the cause of a large amount of the heat which the surface of the earth possesses. On the other hand, the earth in some way parts with a large amount of heat. It is quite apparent that the earth does not receive heat from the sun by conduction or convection, and that by neither of these processes does it part with heat. It is also apparent that there is another and a much more rapid and effectual method by which bodies of higher temperature on the earth part with their heat, and other bodies of lower temperature acquire heat at the expense of distant bodies, than by either of the two comparatively slow processes of diffusion so far described. This process is called

radiation. The process is a very peculiar one, and must be reserved for discussion in its proper place in the chapter on Radiant Energy.

QUESTIONS.

1. Why does more heat reach your hand above than at an equal distance beside a flame?

2. Why is loose clothing warmer than tight-fitting clothing?

3. (*a*) Which contains more heat, the Atlantic Ocean or a tea-kettle full of boiling water? (*b*) Which is capable of giving heat to the other? (*c*) Can a body have less heat than another and yet be hotter than the other?

4. Why should heat be applied to the bottom of a body of water?

5. (*a*) How is equalization of temperature effected in solids? (*b*) In liquids and gases?

Section III.

EFFECTS OF HEAT. — EXPANSION.

105. Expansion of Solids, Liquids, and Gases.

Experiment 91. — The brass ring and ball (Fig. 121) are so constructed that the latter will just pass through the former when both have the same, or nearly the same, temperature. Heat the ball quite hot in a flame, and ascertain by trying to pass it through the ring whether it has increased in size. Devise some method of passing it through the ring without cooling the ball.

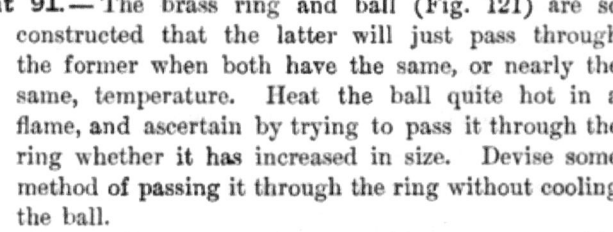

Fig. 121.

Experiment 92. — Figure 122 represents a thin brass plate and an iron plate of the same dimensions riveted together so far as to form what is called a *compound bar*. Place the bar edgewise in a flame, dividing the flame in halves (one-

EFFECTS OF HEAT. — EXPANSION.

half on each side of the bar) so that both metals may be equally heated. The bar, which was at first straight, is now bent, owing to the *unequal expansion* of the two metals on receiving *equal increments of heat*. Which metal expands more rapidly? Thrust the hot bar into cold water. What happens? Cover the bar with chips of ice. What happens?

Experiment 93. — Fit stoppers (perforated rubber stoppers are best) tightly in the necks of two similar thin glass flasks (or test-tubes), and through each stopper pass a glass tube about 18 inches long. The flasks should be nearly of the same size. Fill one flask with water and the other with alcohol, and crowd in the stoppers so as to force the liquids up the tubes a little way above the stoppers. Set both flasks at the same time into a large basin of hot water in order that both may have the same opportunity to acquire heat. Soon the liquids begin to expand and rise in the tubes. Which liquid is more expansible?

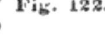

Fig. 122.

Experiment 94. — Take a dry flask like that used in Experiment 94, insert the end of the tube in a bottle of colored water (Fig. 123), and apply heat to the flask; the enclosed air expands and comes out through the liquid in bubbles. After a few minutes, withdraw the heat, keeping the end of the tube in the liquid; as the air left in the flask cools, it loses some of its tension, and the water is forced by atmospheric pressure up the tube into the flask, and partially fills it.

Experiment 95. — Partly fill a foot-ball (see Fig. 9, page 8) with cold air, close the orifice, and place it near a fire. The air will expand and distend the ball.

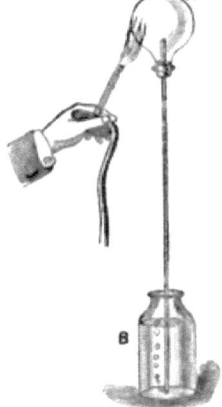

Fig. 123.

Different substances, both in the solid and liquid states, expand unequally on experiencing equal changes of temperature. Except at very low temperatures, *all gases expand alike for equal changes of temperature.* Under uniform pressure (as is very nearly the case in the experiment with the balloon) the volume of any body of gas varies $\frac{1}{273}$ its volume at the freezing-point of water for every degree Centigrade, or [1] for every degree Fahrenheit, its temperature is changed. But if the gas is confined in a vessel of rigid sides, so that its volume is necessarily constant, then its *tension* varies in the same ratio for every degree its temperature is changed.

The force exerted by bodies in expanding or contracting is very great, as shown by the following rough calculation: If an iron bar, 1 square inch in section, is raised from 0° C. (freezing-point of water) to 500° C. (a dull, red heat), its length, if allowed to expand freely, will be increased from 1 to 1.006. Now, a force capable of stretching a bar of iron of 1 square inch section this amount is about 90 tons, which represents very nearly the force that would be necessary to prevent the expansion caused by heat. It would require an equal force to prevent the same amount of contraction if the bar is cooled from 500° to 0° C.

Boiler plates are riveted with red-hot rivets, which, on cooling, draw the plates together so as to form very tight joints. Tires are fitted on carriage-wheels when red hot, and, on cooling, grip them with very great force.

106. Abnormal Expansion and Contraction of Water. — Water presents a partial exception to the general rule that matter expands on receiving heat and contracts on losing it. If a quantity of water at 0° C., or 32° F., is heated, it contracts as its temperature rises, until it reaches 4° C., or about 39° F., when its volume is least, and therefore it has its *maximum density*. If heated beyond this temperature, it expands, and at about 8° C. its volume is the same as at 0°. On cooling, water reaches its maximum density at 4° C., and expands as the temperature falls below that point.

Section IV.

THERMOMETRY.

A thermometer primarily indicates changes in volume: but as changes of volume are caused by changes of temperature, it is commonly used for the more important purpose of indicating *temperature.*

107. Construction of a Thermometer. — A thermometer generally consists of a glass tube of capillary bore, terminating at one end in a bulb. The bulb and part of the tube are filled with mercury, and the space in the tube above the mercury is usually a vacuum. On the tube, or on a plate behind the tube, is a scale to show the hight of the mercurial column.

108. Standard Temperatures. — That a thermometer may indicate any definite temperature, it is necessary that its scale should relate to some definite and unchangeable points of temperature. Fortunately nature furnishes us with two convenient standards. It is found that under ordinary atmospheric pressure ice always melts at the same temperature, called the *melting-point*, or, more commonly, *the freezing-point* (water freezes and ice melts at the same temperature). Again, the temperature of steam rising from boiling water under the same pressure is always the same.

109. Graduation of Thermometers. — The bulb of a thermometer is first placed in melting ice (Fig. 124), and allowed to stand until the surface of the mercury becomes stationary, and a mark is made upon the stem at that point, and indicates the *freezing-point*. Then the instrument is suspended in steam rising from boiling water (Fig. 125), so that all but the very top of the column is in the steam. The mercury rises in the stem until its temperature becomes the same as that of the steam, when it again becomes stationary, and another mark is placed upon the stem to indicate the *boiling-point*. Then the space be-

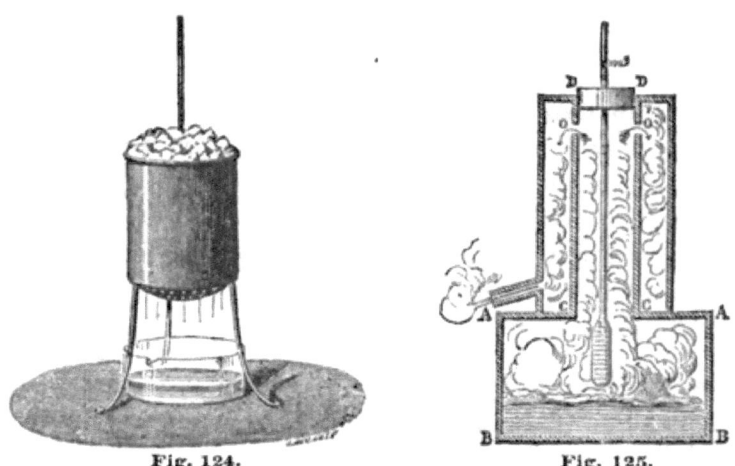

Fig. 124.   Fig. 125.

tween the two points found is divided into a convenient number of equal parts called *degrees*, and the scale is extended above and below these points as far as desirable.

Two methods of division are adopted in this country: by one, this space is divided into 180 equal parts, and the result is called the *Fahrenheit* scale, from the name of its author; by the other, the space is divided into 100 equal parts, and the resulting scale is called *centigrade*, which means *one hundred steps*. In the Fahrenheit scale, which is generally employed in English-speaking countries for ordinary household purposes, the freezing and boiling points are marked respectively 32° and 212°. The 0 of this scale (32° below freezing-point), which is about the lowest temperature that can be obtained by a mixture of snow and salt, was incorrectly supposed to be the lowest temperature attainable. The centigrade scale, which is generally employed by scientists, has its freezing and boiling points more conveniently marked, respectively 0° and 100°. A temperature below 0° in either scale is indicated by a minus sign before the number. Thus, $-12°$ F. indicates 12° below 0° (or 44° below freezing-point), according to the Fahrenheit scale.

To reduce a Fahrenheit reading to a centigrade reading, *first subtract* 32 *from the given number, and then multiply by* $\frac{5}{9}$. Thus,

$$\tfrac{5}{9}(F - 32) = C.$$

To change a centigrade reading to a Fahrenheit reading, *first multiply the given number by* $\frac{9}{5}$, *and then add* 32. Thus,

$$\tfrac{9}{5} C + 32 = F.$$

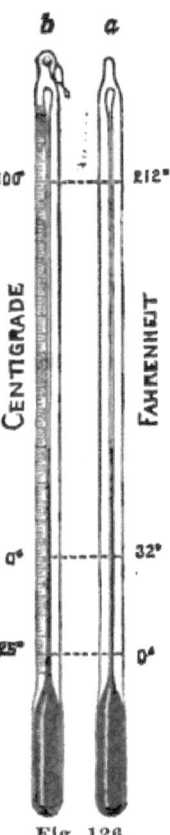

Fig. 126.

EXERCISES.

1. Express the following temperatures of the centigrade scale in the Fahrenheit scale: 100°; 40°; 56°; 60°; 0°; $-20°$; $-40°$; 80°; 150.

NOTE.— In adding or subtracting 32°, it should be done *algebraically*. Thus, to change $-14°$ C. to its equivalent on the Fahrenheit scale: $\tfrac{9}{5} \times (-14) = -25.2°$; $-25.2° + 32° = 6.8°$, the required temperature on the Fahrenheit scale. Again, to find the equivalent of 24° F. in the centigrade scale: $24 - 32 = -8$; $-8 \times \tfrac{5}{9} = -4\tfrac{4}{9}$; hence, 24° F. is equivalent to $-4.4° + $ C.

2. Express the following temperatures of the Fahrenheit scale in the centigrade scale: 212°; 32°; 90°; 77°; 20°; 10°; $-10°$; $-20°$; $-40°$; 40°; 59°; 329°.

Section V.

EFFECTS OF HEAT CONTINUED. — LIQUEFACTION AND VAPORIZATION.

110. Liquefaction. — As previously stated (page 9), whether a body exist in a solid, liquid, or gaseous state depends upon its temperature and the pressure which it is under.

Experiment 96. — Take a lump of ice as large as your two fists, put it into boiling water; when reduced to about ¼ its original size skim it out. Wipe the lump, and place one hand on it and the other on a lump to which heat has not been applied. Do you perceive any difference in their temperatures? Ice reduces the temperature of victuals in our refrigerators; do the victuals raise the temperature of the ice? How does the heat which the victuals impart to the ice affect it?

Experiments and experience teach that (1) *the melting or solidifying point* (they are always the same for the same substance) *may vary widely for different substances, but for the same substance it is invariable when under the same pressure.*

(2) *The temperature of a solid or liquid remains constant at the melting-point from the moment that melting or solidification begins.*

Fig. 127.

111. Vaporization.

Experiment 97. — Place a test-tube (Fig. 127), half filled with ether, in a beaker containing water at a temperature of 60° C. Although the temperature of the water is 40° below its boiling-point, it very quickly raises the temperature of the ether sufficiently to cause it to boil violently. Introduce a chemical thermometer[1] into the test-tube, and ascertain the boiling-point of ether.

---

[1] A chemical thermometer has its scale on the glass stem, instead of a plate, and is otherwise adapted to experimental use.

LIQUEFACTION AND VAPORIZATION.   137

Experiment 98. — Take two beakers half full of water. Raise both to the boiling-point. Dissolve pulverized saltpetre in one as long as it readily dissolves. Suspend in both liquids chemical thermometers, so that the bulb of each shall be within one inch of the bottom. Does the boiling water, as you continue to apply heat, get hotter? Is the boiling solution any hotter than the boiling water? Does the solution get hotter as it becomes concentrated by loss of water by vaporization?

After a liquid begins to boil, the temperature remains constant until the whole is vaporized, if the density of the liquid and the pressure remain constant.

Experiment 99. — Place a beaker, half full of water at 80° C., under the receiver of an air-pump, and exhaust the air. The water, though far below its usual boiling-point, boils violently. Readmit the air, and test the temperature of the water which has just been boiling.

Fig. 128.   Fig. 129.

Experiment 100. — Half fill a thin glass flask with water. Boil the water over a Bunsen burner; the steam will drive the air from the flask. Withdraw the burner, quickly cork the flask very tightly, invert the flask, and pour cold water upon the part containing steam, as in Figure 128; the water in the flask, though cooled several degrees below the usual boiling-point, boils again violently. The application of cold water to the flask condenses some of the steam, and diminishes the tension of the rest, so that the pressure upon the water is diminished, and the water boils at a reduced temperature.

If hot water is poured upon the flask, the water ceases to boil. Why?

Experiment 101. — Provide a tumbler of cold water, a test-tube nearly filled with water, tightly stoppered, and having glass tubes extending through the stopper, as represented in Figure 129. Place the exposed end of the bent tube in the tumbler of water, and apply heat to the bottom of the test-tube, and boil the water for about five minutes. Then remove the heat, leave the end of the tube in the tumbler of water, and allow the water of the test-tube to cool for some time; or, better, to hasten the cooling, place the test-tube in another tumbler of cold water. Observe carefully, and explain all phenomena which occur from the beginning to the end of the operation.

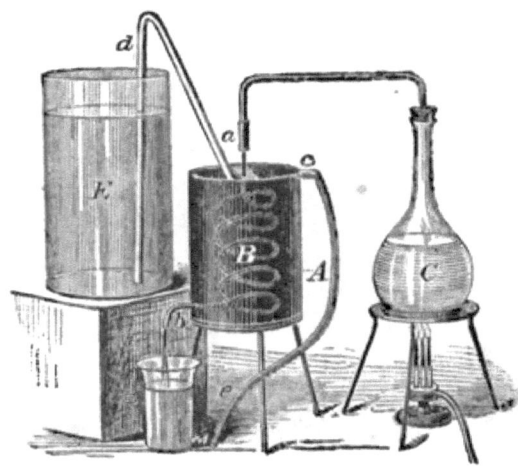

Fig. 130.

112. Distillation.
Experiment 102. — Vessel A (Fig. 130) (called a *condenser*) contains a coil (called a *worm*) of copper tube, terminating at one extremity at *a*. The other end of the tube, *b*, projects through the side of the vessel near its bottom. Near the top of the vessel projects another tube, *c* (called the *overflow*), with which is connected a rubber tube, *e*. This tube conveys the warm water which rises from the surface of the heated worm away to a sink or other convenient receptacle.

Take a glass flask of a quart capacity, fill it three-fourths full of pond or bog water. Connect the flask by means of a glass delivery-tube with the extremity *a* of the worm. Heat the water in the flask; as soon as it begins to boil, commence siphoning cold water through a small tube, *d*, from an elevated vessel E into the condenser. Inasmuch as the worm is constantly surrounded with cold water, the steam on passing through it becomes condensed into a liquid, and the liquid (called the *distillate*) trickles from the extremity *b* into a receiving vessel. The distillate is clear, but the water in the flask acquires a yellowish brown tinge as the boiling progresses, due to the concentration of impurities (largely of vegetable matter) which are held in suspension and solution in ordinary pond water. The apparatus used is called a *still*, and the operation *distillation*.

When a volatile liquid is to be separated from water, for example, when alcohol is separated from the vinous mash after fermentation (see Chemistry, page 184), the mixed liquid is heated to its boiling-point, which is lower than that of water. Much more of the volatile liquid will be converted into vapor than of the water, because its boiling point is lower. Thus a partial separation is effected. By repeated distillations of the distillate, a 95 per cent alcohol is obtained.

113. Evaporation. — In boiling, the heat, applied at the bottom, rapidly converts the liquid into vapor, which, rising in bubbles and breaking at or near the surface, produces a violent agitation in the liquid, called boiling or *ebullition*. Boiling takes place only at a definite temperature, which depends on the kind of liquid and the pressure that is on it. *Evaporation* is that form of vaporization which takes place quietly and slowly at the surface. Although hastened by heat, the evaporation of water occurs at *any temperature*, however low; even ice and snow evaporate.

The rapidity of evaporation increases with the temperature, amount of surface exposed, dryness of the atmosphere, and diminution of pressure. This vapor of water mixes freely with the air, and diffuses rapidly through it, acting like another gas. A given space, — for example, a cubic foot (it matters little whether there is air in the space or whether it is a vacuum), can hold only a limited amount of water vapor. This quantity depends on the temperature of the vapor. The capacity of a space for water vapor increases rapidly with the temperature, being nearly doubled by a rise of 10° C. When a space contains such an amount of water vapor that its temperature cannot be lowered without some of the water being precipitated in the form of a liquid, the vapor is said to be *saturated*, and the temperature at which this happens is called the *dew-point*.

Experiment 103. — Take a bright nickel-plated cup, such, for example, as are used for lemonade-shakers; pour into it a small quantity of tepid water. Place in the water the bulb of a chemical thermometer. Gradually reduce the temperature of the water by stirring into it ice water until you discover a slight dimness of the luster of that portion of the outside of the cup next the water. If the ice water does not reduce the temperature sufficiently, add ice, keeping the mixture briskly stirring. If the ice does not answer, pour out some of the water and sprinkle salt on the ice, keeping the bulb of the thermometer in the remaining water. Note the temperature of the water at the instant that the first mist or dimness appears on the cup. Wait until the dimness or mist disappears, and note the temperature of the water when the last disappears. Take the mean of the two temperatures for the dew-point.

The form in which the condensed vapor appears is, according to its location, *dew*, *fog*, or *cloud*.[1] The atmosphere is said to be *dry* or *humid*, not according to the quantity of water vapor which it at any time contains, but according as it can contain much or little more than it has. The air in summer months usually contains a large amount of water vapor, yet it is usually very dry. The heat of a stove *dries* the air of a room without destroying any of its water vapor. In such a room, the lips, tongue, throat, and skin experience a disagreeable sensation of dryness, owing to the rapid evaporation which takes place from their surfaces. This should be taken as nature's admonition to keep water in the stove urns, and tanks connected with furnaces.

---

[1] A cloud is simply a fog in an elevated region of the atmosphere. It is composed of minute spheres of water from $\frac{1}{1000}$ to $\frac{1}{1000}$ of an inch in diameter.

Section VI.

HEAT CONVERTIBLE INTO POTENTIAL ENERGY, AND VICE VERSA.

114. Heat Units. — It is frequently necessary to measure *quantity of heat*, and for this purpose a standard unit of measurement is required. The heat unit generally adopted is *the amount of heat required to raise the temperature of one kilogram of water from* $0°$ *to* $1°$ C. This unit is called a *calorie*.

Let it be required to find approximately the amount of heat that disappears during the melting of one kilogram of ice.

Experiment 104. — Weigh out 200^g of dry (dry it with a towel) ice chips whose temperature in a room of ordinary temperature may be safely assumed to be $0°$ C. Weigh out 200^g of boiling water, whose temperature we assume to be $100°$ C. Pour the hot water upon the ice, and stir until the ice is all melted. Test the temperature of the resulting liquid.

Suppose its temperature is found to be $10°$ C. It is evident that the temperature of the hot water in falling from $100°$ to $90°$ would yield sufficient heat to raise an equal weight of water from $0°$ to $10°$ C. Hence it is clear that the heat which the water at $90°$ yields in falling from $90°$ to $10°$ — a fall of $80°$ — in some manner disappears. At this rate had you used 1^k of ice and 1^k of hot water, the amount of heat lost would be 80 calories. Careful experiments, in which suitable allowances are made for loss or gain of heat by radiation and conduction, have determined that 80 *calories of heat are consumed in melting* 1 *kilogram of ice.* How near to this do the results of your experiments approach?

Next let it be required to find the amount of heat that disappears during the conversion of 1 kilogram of water into steam.

Experiment 105. — Take in a porcelain evaporating-dish 50^g of ice water at (say) 5° C. Place it over a flame, and, watch in hand, note the time in seconds which elapses before it boils. Then note the time which elapses before it is all converted into steam. Suppose that it required 100 seconds to raise the water from 5° to its boiling-point, which we assume is 100° — a rise of 95°; and that it requires 530 seconds to convert the water, after it commences to boil, into steam. Then the latter operation consumes (530÷100=) about 5.3 times as much time as the former. But the heat applied to the water while boiling does not raise its temperature (see Exp. 98, page 137); then all the heat given to the water during the interval of time disappears. Had you taken 1^k of water, it would have required 95 calories to raise the water from 5° to 100° C. Hence, in converting the 1^k of water into steam, $95 \times 5.3 =$ (about) 503 calories disappear. Accurate methods have determined that 537 calories disappear during the conversion of 1^k of water into steam.

The heat which disappears in melting and boiling is generally, but with our present knowledge of the subject, rather objectionably, called *latent* (hidden) *heat*. The error consists in calling that heat which has ceased to be heat. The heat, *i.e.* kinetic energy, that disappears in melting is consumed in doing *interior* (*i.e.* molecular) *work*. The molecules that in the solid are held firmly in their places by molecular forces, are moved from their places during melting, and so work is done against these forces, much as work is done against gravity when a stone is raised. In both cases *kinetic* energy is consumed — disappears; but this means simply that it is transformed into *potential* energy. The so-called latent heat is simply a misnomer for *molecular potential energy*.

When water is converted into steam, the larger portion of the heat, which is rendered latent, is consumed in separating the molecules so far that molecular attraction is no longer sensible; a small portion — about $\frac{1}{13}$ — is consumed in overcoming atmospheric pressure. The amount of work done in melting and boiling — especially the latter — is very great, as shown by the amount of heat consumed. Hence it requires a long time to acquire the requisite amount of heat. This is a protection against sudden changes. For example, if snow and ice melted immediately on reaching the melting-point, all the snow and ice would melt in a single warm day in winter, creating most destructive freshets.

115. Potential Energy converted into Heat by the Solidification of Liquids and the Liquefaction of Vapors. — If our theory be true that heat is converted into potential energy during vaporization and melting, then ought the energy to be restored to the kinetic state (*i.e.* the heat which disappears during these operations ought to be restored) when the molecules return to their original positions, *i.e.* when vapor becomes liquid, or when liquids solidify.

Experiment 106. — Take in a beaker C (Fig. 131) 1^k of water at (say) 12° C. Take about 500^g of water in a flask A, and raise it to the boiling-point. As soon as it begins to boil, connect the flask with the beaker by a delivery-tube B, carrying the end of the tube nearly to the bottom of the beaker. When about one-fifth of the water has boiled away, remove the delivery tube from C, and immediately test the temperature of the water in the beaker, and weigh it. Assume that the temperature of the steam is 100° C., and we will suppose, for illustration, that there are $1,100^g$ of water now in the beaker; then 100^g of water have been converted into steam which has passed into the beaker and been condensed or liquefied by the cold water. Suppose, again, that the temperature of the water in the beaker was raised thereby to 70° C. Now 100^g of water at 100° C. (resulting from the condensation of the steam) in falling to 70° C. could yield (30+10=) only 3 calories; hence it could raise the 1^k of water only 3°; *i.e.* from 12° to 15° C. Then it is evident that it must have acquired the balance of (70 − 15 =) 55 calories, by the

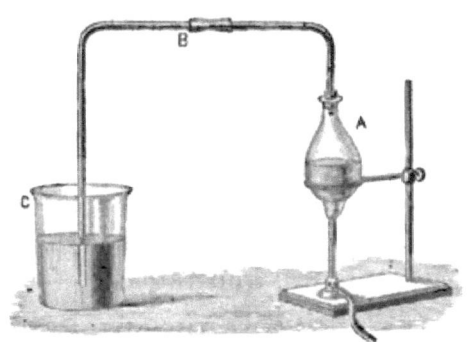

Fig. 131.

restoration of the latent heat to *real* heat when the steam is liquefied. If the liquefaction of 100^g of steam yields 55 calories, then the liquefaction of 1^k of steam would yield 550 calories. Accurate methods give 537 calories.

Various phenomena show that heat is developed during the solidification of liquids, but as the development is slow, and the loss by radiation rapid, it is difficult to make measurements. There are good reasons for assuming, however, that there are 80 calories of heat generated for every kilogram of water that is frozen. Farmers sometimes turn to practical use this well-known phenomenon. Anticipating a cold night, they carry tubs of water into cellars to be frozen. The heat generated thereby, although of a low temperature, is sufficient to protect vegetables which freeze at a lower temperature than water.

Steam is a most convenient vehicle for the conveyance of latent heat. For example, every kilogram of steam that is condensed in the radiator box (Fig. 120, p. 128) contributes to the air which passes through the box 537 calories, or heat sufficient to raise 5.37^k of ice water to the boiling-point.

116. Methods of Producing Artificial Cold.

The fact that heat must be consumed because work is done, in the conversion of solids into liquids and liquids into vapors, is turned to practical use in many ways for the purpose of producing *artificial cold*. The following experiments will illustrate.

117. Cold by Dissolving. — Freezing Mixtures.

Experiment 107. — Prepare a mixture of 2 parts, by weight, of pulverized ammonium nitrate and 1 part of ammonium chloride. Take about 75^{cc} of water (not warmer than 8° C.), and into it pour a large quantity of the mixture, stirring the same, while dissolving, with a test-tube containing a little cold water. The water in the test-tube will be quickly frozen. A finger placed in the solution will feel a painful sensation of cold, and a thermometer will indicate a temperature of about $-10°$ C.

One of the most common freezing mixtures consists of 3 parts of snow or broken ice and 1 part of common salt. The affinity of salt for water causes a liquefaction of the ice, and the resulting liquid dissolves the salt, both operations requiring heat.

118. Cold by Evaporation.

Experiment 108. — Fill the palm of the hand with ether; the ether quickly evaporates, and produces a painful sensation of cold.

Experiment 109. — Place water at about 30° C. in a thin porous cup, such as is used in the Grove's battery, and the same amount of water, at the same temperature, in a glass beaker of as nearly as possible the same size as the porous cup. Introduce into each a chemical thermometer. The comparatively large amount of surface exposed by means of the porous vessel will so hasten the evaporation in this vessel, that, in the course of 10 to 15 minutes, quite a sensible difference of temperature will be indicated by the thermometers in the two vessels.

Experiment 110. — Cover closely the bulb of an air thermometer (Fig. 132) with thin muslin, and partly fill the stem with water. Let one person slowly drop ether on the bulb, while another briskly blows the air charged with vapor away from the bulb with a bellows; or, place the bulb in a window whose sash is raised a little way, so as to be in a draft. As the air changes rapidly, it does not become saturated with vapor so as to impede evaporation, and in 10 to 15 minutes the water in the stem freezes, even in a warm room.

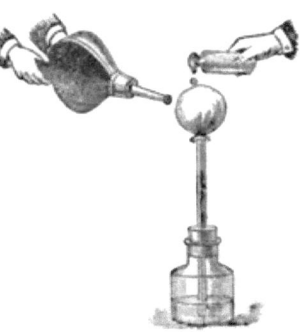

Fig. 132.

The evaporation of perspiration conduces to our health and comfort by relieving us of surplus heat. We cool the fevered forehead by bathing it with a volatile liquid, such as a solution of alcohol in water. Windy days seem colder to us than still days, although the temperature of both is the same, because evaporation of perspiration takes place more rapidly in a changing air. Fanning in a similar way changes the air next our persons, thereby quickening the evaporation of the perspiration, and cooling the surface of the body. Ice is now manufactured in large quantities during the summer season in warm climates by the evaporation of liquid ammonia. Evaporation is the most efficient means of producing extremely low temperatures.

QUESTIONS.

1. How can water be made to boil at a low temperature?
2. Upon what does the tension of steam depend?
2. Why can you not make ice warm?
4. Does ice always have the same temperature; *i.e.* can it be made colder than 32° F.?
5. What is the lowest temperature any body can have?
6. (*a*) Where does the "sweat" on ice-pitchers come from? (*b*) Where does dew on grass come from? (*c*) How are clouds formed?
7. (*a*) When the sweat on ice-pitchers is very abundant, what does it indicate about dew-point? (*b*) Does it forebode fair or rainy weather?
8. How will you easily show that ether boils at a lower temperature than water?
9. In which will vegetables cook quicker, — in fresh or salt water?
10. How could you separate the alcohol of rum or brandy from the watery part?
11. (*a*) On what kind of days do clothes dry fastest? (*b*) Will frozen clothes dry?
12. (*a*) How does heat dry the air? (*b*) How does heat dry clothes?
13. Suppose that 10^k of steam, at 100° C., is condensed in the steam-pipe in the radiator box, Figure 120, per hour; how much heat will it furnish to the surrounding air?
14. How much heat will be produced by freezing one cubic foot (about 29^k) 62.5 pounds of water?

Section VII.

THERMO-DYNAMICS.

119. Thermo-dynamics Defined. — *Thermo-dynamics is that branch of science that treats of the relation between heat and mechanical work.* One of the most important discoveries in science is that of the *equivalence of heat and work;* that is, that *a definite quantity of mechanical work, when transformed without waste, will yield a definite quantity of heat; and conversely, this heat, if there were no waste, could perform the original quantity of mechanical work.*

120. Transformation, Correlation, and Conservation of Energy. — The proof of the facts just stated was one of the most important steps in the establishment of the grand twin conceptions of modern science: (1) That *all kinds of energy are so related to one another that energy of any kind can be transformed into energy of any other kind,* — known as the doctrine of CORRELATION OF ENERGY; (2) That *when one form of energy disappears, an exact equivalent of another form always takes its place, so that the sum total of energy is unchanged,* — known as the doctrine of CONSERVATION OF ENERGY. These two principles constitute the corner-stone of physical science. Chemistry teaches that there is a conservation of matter.

121. Joule's Experiment. — The experiment to ascertain the "mechanical value of heat," as performed by Dr. Joule of England, was conducted about as follows. He caused a paddle-wheel to revolve in water, by means of a falling weight attached to a cord wound around the axle of a wheel. The resistance offered by the water to the motion of the paddles was the means by which the mechanical energy of the weight was converted into heat, which raised the temperature of the water. Taking a body of a known weight, *e.g.* 80^k, he raised it a measured distance, *e.g.* 53^m high; by so doing $4,240^{kgm}$ of work were performed upon it, and consequently an equivalent amount of energy was stored up in it ready to be converted, first into mechanical motion, then into heat. He took a definite weight of water to be agitated, *e.g.* 2^k, at a temperature of $0°$ C. After the descent of the weight, the water was found to have a temperature of $5°$ C.; consequently the 2^k of water must have received 10 units of heat (careful allowance being made for all losses of heat), which is the amount of heat-energy that is equivalent to $4,240^{kgm}$ of work, *or one unit of heat is equivalent to 424^{kgm} of work.*

122. Mechanical Equivalent of Heat. — As a converse of the above it may be demonstrated by actual experiment that the quantity of heat required to raise 1 of water from $0°$ to $1°$ C. will, if converted into work, raise a 424^k weight 1^m high, or 1^k weight 424^m high. According to the English system, the same fact is stated as follows: The quantity of heat that will raise 1 pound of water $1°$ F. will raise 772.55 pounds 1 foot high. The quantity, 424^{kgm}, is called the *mechanical equivalent of one calorie, or Joule's equivalent* (abbreviated simply J.). Or, we may say that one calorie is the *thermal equivalent* of 424^{kgm} of work.

Section VIII.

STEAM-ENGINE.

123. Description of a Steam-Engine. — A steam-engine is a machine in which the elastic force of steam is the motive power. Inasmuch as the elastic force of steam is entirely due to heat, *the steam-engine is properly a heat engine;* that is, it is a machine by means of which heat is continuously transformed into work or mechanical motion.

The modern steam-engine consists essentially of an arrangement by which steam from a boiler is conducted to both sides of a piston alternately; and then, having done its work in driving the piston to and fro, is discharged from both sides alternately, either into the air or into a condenser. The diagram in Figure 133 will serve to illustrate the general features and the operation of a steam-engine. The details of the various mechanical contrivances are purposely omitted, so as to present the engine as nearly as possible in its simplicity.

In the diagram, B represents the *boiler*, F the *furnace*, S the *steam-pipe* through which steam passes from the boiler to a small chamber VC, called the *valve-chest*. In this chamber is a *slide-valve* V, which, as it is moved to and fro, opens and closes alternately the passages M and N leading from the valve-chest to the *cylinder* C, and thus admits the steam alternately each side of the piston P. When one of these passages is open, the other is always closed. Though the passage between the valve-chest and the space in the cylinder on one side of the piston is closed, thereby preventing the entrance of steam into this space, the passage leading from the same space is open through the interior of the valve, so that steam can escape from this space through the *exhaust-pipe* E. Thus, in the position of the valve represented in the diagram, the passage N is open, and steam entering the cylinder at the top drives the piston in the direction indicated by the arrow. At the same time the steam on the other side of the piston escapes through the passage M and the exhaust-pipe E. While the piston moves to the left, the valve moves to the

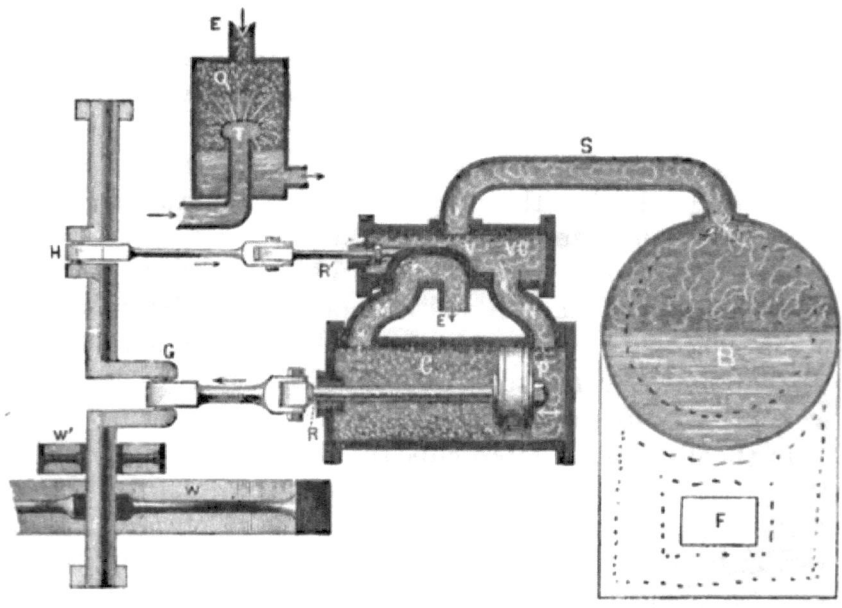

Fig. 133.

right, and eventually closes the passage N leading from the valve-chest, opens the passage M into the same, and thus the order of things is reversed.

Motion is communicated by the piston through the *piston-rod* R to the *crank* G, and by this means the *shaft* A is rotated. Connected with the shaft by means of the crank H is a rod R' which connects with the valve V, so that, as the shaft rotates, the valve is made to slide to and fro, and always in the opposite direction to that of the motion of the piston.

The shaft carries a *fly-wheel* W. This is a large, heavy wheel, having the larger portion of its weight located near its circumference; it serves as a reservoir of energy which is needed to make the rotation of the shaft and all other machinery connected with it uniform, so that sudden changes of velocity resulting from sudden changes of the driving power or resistances are avoided. By means of a belt passing over the wheel W' motion may be communicated from the shaft to any machinery desirable.

124. Condensing and Non-Condensing Engines.[1] — Sometimes steam, after it has done its work in the cylinder, is conducted through the exhaust-pipe to a chamber Q, called a *condenser*, where, by means of a spray of cold water introduced through a pipe T, it is suddenly condensed. This water must be pumped out of the condenser by a special pump, called technically the *air-pump;* thus a partial vacuum is maintained. Such an engine is called a *condensing engine*. The advantage of such an engine is obvious, for if the exhaust-pipe, instead of opening into a condenser, communicates with the outside air, as in the *non-condensing engine*, the steam is obliged to move the piston constantly against a resistance arising from atmospheric pressure of 15 pounds for every square inch of the surface of the piston. But in the condensing engine no resistance arises from atmospheric pressure, and so with a given steam pressure in the boiler the effective pressure on the piston is considerably increased; hence, condensing engines are usually more economical in their working.

---

[1] The terms, *low-pressure* and *high-pressure* engines, are not distinctive as applied to engines of the present day.

125. Compound Condensing Engine. — This engine has two cylinders, each like that of a simple engine. One, A (Fig. 134), called the *high-pressure cylinder*, receives steam of very high pressure directly from the boiler. The steam, after it has done work in this cylinder, passes through the steam-port into cylinder B, *called the low-pressure cylinder*. Cylinder B is larger than cylinder A. The steam which enters cylinder B possesses considerable tension, and is therefore capable of doing considerable work under suitable conditions. It should be borne in mind that in order that steam may do work in any cylinder, it is necessary

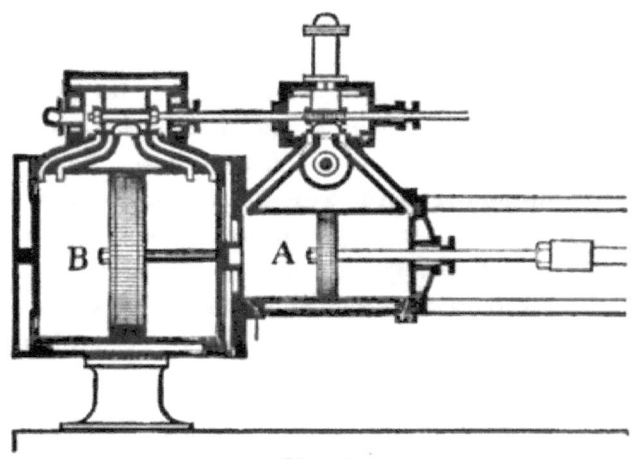

Fig. 134.

that an inequality in the tension of the steam each side of the piston should be maintained; just as an inequality of level, *i.e.* a *head*, is essential to water-power. The steam, after it has done its work in cylinder B, passes through a port into a condenser (not represented in the figure), where it is suddenly condensed or let down to a very low tension. If a vertical glass tube were led from the condenser to a vessel of mercury below, the mercury would ordinarily stand about 25 inches high in the tube, which would show that the tension of the steam against which the steam when it enters cylinder B does work, is only about one-sixth of an atmosphere. Much energy is economized by the compound engine.

126. The Locomotive. — The distinctive feature of the locomotive engine is its great steam-generating capacity, considering its size and weight, which are necessarily limited. To do the work ordinarily required of it, from three to six tons of water must be converted into

STEAM LOCOMOTIVE.   JUNE '05.

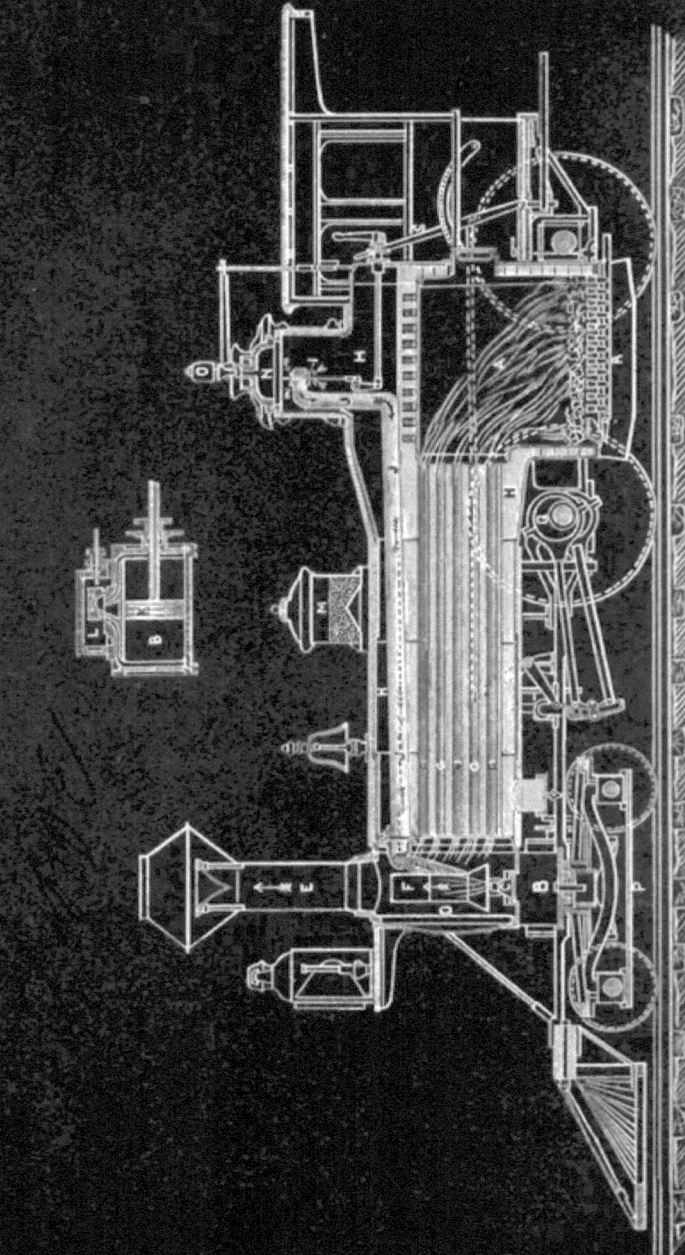

steam per hour. This is accomplished in two ways: viz., first, by a rapid combustion of fuel (from a quarter of a ton to a ton of coal per hour); second, by bringing the water in contact with a large extent (about 800 square feet) of heated surface. The fire in the "fire-box" A (Fig. 135, Plate II.) is made to burn briskly by means of a powerful draft which is created in the following manner: The exhaust steam, after it has done its work in the cylinders B, is conducted by the exhaust-pipe C to the smoke-box D, just beneath the smoke-stack E. The steam, as it escapes from the blast-pipe F, pushes the air above it, and drags by friction the air around it, and thus produces a partial vacuum in the smoke-box. The external pressure of the atmosphere then forces the air through the furnace grate and hot-air tubes G, and thus causes a constant draft. The large extent of heated surface is secured as follows: The water of the boiler is brought not only in contact with the heated surface of the fire-box, but it surrounds the pipes G (a boiler usually contains about 150). These pipes are kept hot by the heated gases and smoke, all of which must pass through them to the smoke-box and smoke-stack.

The steam-engine, with all its merits and with all the improvements which modern mechanical art has devised, is an exceedingly wasteful machine. The best engine that has been constructed utilizes only about twenty per cent of the heat-power generated by the combustion of the fuel.

QUESTIONS.

1. What kind of engine (*i.e.* condensing or non-condensing) is that which produces loud puffs? What is the cause of the puffs?
2. Why does the temperature of steam suddenly fall as it moves the piston?
3. What do you understand by a ten horse-power steam-engine?
4. Upon what does the power of a steam-engine depend?
5. Is the compound engine a condensing or a non-condensing engine? Which is the locomotive engine?
6. The area of a piston is 500 square inches, and the average unbalanced steam pressure is 30 pounds per square inch; what is the total effective pressure? Suppose that the piston travels 30 inches at each stroke, and makes 100 strokes per minute, allowing 40 per cent for wasted energy, what power does the engine furnish, estimated in horse-powers?

CHAPTER VI.

ELECTRICITY AND MAGNETISM.

Section I.

INTRODUCTORY EXPERIMENTS.

No other department of Physics presents so many favorable opportunities for individual work as that of Electricity. There is none in connection with which apparatus sufficient to equip a laboratory can be provided so cheaply, when the amount of work which can be done with it is considered; certainly there is no other department in connection with which laboratory work is so *indispensable* in order to acquire a *working* knowledge of the subject.

127. Apparatus Required. — A tumbler $\tfrac{2}{3}$ full of water into which has been poured two or three tablespoonfuls of strong sulphuric acid; a strip of sheet-copper, and two pieces of zinc, each about 5 inches long and $1\tfrac{1}{2}$ inches wide. The pieces of zinc should be $\tfrac{3}{16}$ of an inch thick. A piece of No. 16 copper wire, 12 inches long, should be soldered to one end of each piece of metal. The soldering should be covered with asphaltum paint. Also, a rod of Norway iron, 6 inches long and $\tfrac{1}{4}$ of an inch in diameter; 4 yards of No. 23 insulated copper wire; a magnetic needle, 6 inches long, nicely poised on a fine needle-point; some fine iron turnings; and two double connectors. These connectors (Fig. 136) serve to connect two wires, without the inconvenience of twisting them together. Wind the wire closely, with the exception of about 10 inches at each extremity, around the iron-rod, nearly from end to end, in two or three layers, as the case may require. Amalgamate one of the zincs as follows: first dip the zinc, with the exception of about $\tfrac{1}{2}$ an inch at the soldered end, into the acidulated water; then pour mercury over the surface, and finally rub the surface wet with mercury with a cloth.

Fig. 136.

INTRODUCTORY EXPERIMENTS. 155

Experiment 111. — Put the unamalgamated (dark colored) zinc into the liquid. Bubbles of gas arise from the zinc. These bubbles, Chemistry (page 24) teaches, are hydrogen gas. Put the copper strip into the liquid, but do not allow the two metals or their wires to touch. Do bubbles rise from the copper?

Experiment 112. — Remove the metals, and allow the liquid to become clear. Connect their wires with a double connector, and introduce both metals into the liquid, about 1 inch apart. Hold them perfectly still for a minute, and observe whether any bubbles escape from the copper.

Bubbles escaping from both metals make it appear as if chemical action were taking place between both metals and the liquid. But experience will teach you that the appearance is deceptive, as you will find that only the zinc is consumed.

Experiment 113. — Put the amalgamated (bright) zinc into the liquid. If the zinc is properly amalgamated, no bubbles will rise from it. Do you discover any? If so, report it to your teacher.

Experiment 114. — Put the copper strip into the liquid. Do not allow the metals or their wires to touch. Do bubbles rise from either metal? Connect their wires. Do bubbles now rise from either metal?

Lesson learned: — An amalgamated zinc is not acted on by the liquid unless a copper strip is also in the liquid, and not then unless the metals are connected. If then we would at any time stop the action, we have only to disconnect the metals.

It seems that the wire connector serves a very important purpose. *Does it, meantime, possess any unusual properties?*

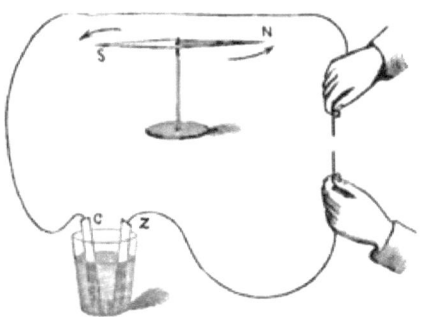

Fig. 137.

Experiment 115. — Place a magnetic needle (Fig. 137) near your tumbler. When the needle comes to rest, it points north and south.

Place the connecting wire over and near the needle, so that that portion of the wire which is over the needle shall have a northerly and southerly direction. Does the needle move? Does the end of the needle pointing to the north (called the *north-seeking pole* of the needle) move toward the east or the west? Imagine that your wire is a tube through which there is flowing a liquid. Turn the tumbler half way around, so that the current in that portion of the wire which is over the needle shall be reversed. Do you observe any change in the deflection of the needle? Next, lower that portion of the wire which is over the needle, and place it nearly under the needle. Do you observe any change in the deflection of the needle?

Lesson learned: — (1) The wire connector does possess an unusual property. It is capable of exercising an unusual form of force. This new form of force is called *electro-magnetic force*. (2) Although we have no *positive* evidence that anything of the nature of a fluid flows through the wire, yet in discussing certain phenomena, such, for example, as the deflection of the needle, it is extremely convenient, at least, to *imagine* that a current passes through the wire. Something *does* pass through the wire. What this something is, physicists have not determined. They have merely given it a name — *electricity*.

128. Some Technical Terms. — Experiments, not easily performed by the pupil, show that the current traverses the liquid between the metals at the same time that it traverses the connecting wire, so that the current makes a complete circuit. The term *circuit* is applied to the entire path along which electricity flows, and the wire through which it flows is called the *conductor*. Bringing the two extremities of the wires, or other parts of the circuit, in contact (so as to complete the circuit) and separating them, are called, respectively, *closing and opening, or making and breaking, the circuit*. Opening a circuit at any point, and filling the gap with an instrument of any kind, so that the current is obliged to pass through it, is called *introducing* an instrument *into the circuit*. Our arrangement of acidulated water and two metals is called a *voltaic cell*, and the two metals are called its *elements*. A series of cells properly connected is called a *battery*, though this term is frequently applied to a single cell. The free extremities of the wires are called *poles* or *electrodes*, and the same terms may be applied to any two points of contact in any part of the circuit.

129. Conductors and Non-Conductors.

Experiment 116. — Will every substance answer for a conductor? Introduce into the circuit between the electrodes, pieces of wood, paper, cloth, glass, iron, brass, zinc, lead; also, a drop of mercury on a glass plate. Place the connecting wire over the magnetic needle, and determine, by the deflection of the needle, through which of these substances electricity will pass. Those substances through which electricity passes readily are called *good conductors*. Substances through which electricity passes with great difficulty are called *bad conductors*, *non-conductors*, or *insulators*. Are metals conductors or non-conductors?

130. Direction of the Current, etc. — It is evidently necessary in describing a current to assign it a direction.

Electricians have universally agreed, for the purpose of uniformity and convenience, to assume that in such a cell as described, electricity flows from the zinc, where the chemical action takes place, through the liquid to the copper element, thence through the wire to the starting-point, *i.e.* the zinc element. If we take any two points in a circuit, of course the current will be *from* one *toward* the other. The former point is said to be *positive* (+) with reference to the latter point which is said to be *negative* (−). Which is the positive element of a battery, the zinc or the copper plate? Which electrode, *i.e.* the free end of the wire connected with the zinc plate, or the end of the wire connected with the copper plate, is positive?

Experiment 117. — Place the connecting-wire over the magnetic needle, in such a manner that the current will flow northward through that section of the wire that is above the needle. Then reverse the direction of the current. Finally, place the wire under the needle. In each different position verify the following rule for determining the direction of the deflection when the direction of the current is known.

131. Ampère's Rule. — Imagine yourself to be swimming in the current, and *with* (*i.e.* your head in the same direction as) the current, and *facing* (*i.e.* looking up or down according as the needle is above or below you) the needle; in such a position the north pole of the needle is always deflected toward your *left*.

Fig. 138.

132. Galvanoscope. — The magnetic needle serves the purpose of determining the presence of a current in a wire. A needle used for this purpose is called a *galvanoscope*. Electricity set in motion by a voltaic battery is called *galvanic* or *voltaic* and sometimes *current* or *dynamic electricity*.

Section II.

POTENTIAL AND ELECTRO-MOTIVE FORCE.

133. Potential. — In order that water may flow from one vessel A to another B through a connecting pipe, there must be a difference of level in the two vessels; *i.e.* in ordinary language there must be a *head* of water in A. The head of water in A causes a greater *pressure* at the end of the pipe next this vessel than at the end next B, and this unbalanced force causes a flow of water from A to B until there is the same level in both vessels. So, in order that there may be a flow of electricity from a body A to a body B, or from a given point A in a body to another given point B in the same body, there must be a difference of *electrical condition* between A and B. This difference of condition may be imagined as a difference of electrical pressure and is called a *difference of electrical potential.*

In any case we may say that difference of potential with reference to electricity is analogous to difference of pressure in fluids, and that electricity always tends to flow from places of high to places of low potential.

134. Electro-Motive Force. — When two conductors are connected by a wire it is found that the rate at which electricity passes from one to the other is proportional to the difference of potential of the two conductors, and that this is proportional to the work that would be expended in carrying a unit quantity of electricity backwards through the wire. So, too, in any circuit it is found that the quantity of electricity flowing in any time is strictly proportional to the amount of work necessary to carry a unit quantity of electricity backwards through the circuit. This important magnitude receives the name *electro-motive force* (E.M.F.), although it is apparent that it is *not* a force. *The E.M.F., or work done, is the cause of difference of potential.*

We might agree to call any point in a liquid stream positive with reference to all points below it or of lower level, and negative with reference to all points above it, or points of higher level. So *any point in an electrical conductor is said to be positive with reference to all points of lower potential, and negative with reference to all points of higher potential.*

Is there such a thing as a difference of electrical condition?

Fig. 139.

Experiment 118. — Separate a little way the two conductors C and D (Fig. 139), of a Holtz machine. Hold two pith balls suspended by white silk threads against one of the conductors and turn the plate a few times. Remove the pith balls and hold them near each other. They repel each other. Next place one of the pith balls in contact with one of the conductors and the other pith ball in contact with the other conductor. Turn the plate as before, remove the pith balls, and hold them near each other. Now, in the first case, the two pith balls were placed in contact with the same conductor, and hence acquired the same electrical condition (*i.e.* potential) as that conductor. In this condition they repel each other. But after being in contact with different conductors, as in the second case, they attract each other (Fig. 140). Hence, we conclude that the two conductors are in a different electrical condition.

It is in consequence of this difference of electrical condition, which always exists between such bodies, that the electricity passes from one conductor through the air to the other, rendering the air in its path temporarily luminous. The most convenient test of the electro-motive force of an electrical machine is the length of spark it gives.

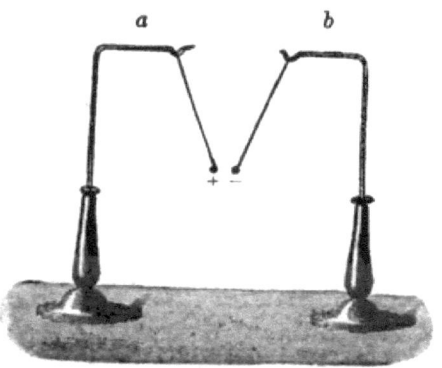

Fig. 140.

135. Electro-Chemical Series.

Experiment 119. — Take two plates of zinc, either both amalgamated or both unamalgamated, connect them, put them into acidulated water, and place the connecting wire over a magnetic needle. Does the galvanoscope show that there is a current in the wire? Is there, then, a difference of potential between the two plates?

It is important that *only one of the two elements of a voltaic cell should be acted upon by the liquid. The greater the disparity between the two solid elements, with reference to the action of the liquid on them, the greater the difference in potential; hence, the greater the current.* In the following *electro-chemical series* the substances are so arranged that the electro-positive, or those most affected by dilute sulphuric acid, are at the beginning; while the electro-negative, or those least affected by the acid, are at the end. The arrow indicates the direction of the current through the liquid.

$$+ \text{ Zinc. Iron. Tin. Lead. Copper. Silver. Platinum. Carbon. } -$$

It will be seen that zinc and carbon are the two substances best adapted to give a strong current.

The essential parts of a galvanic cell are a liquid and two different conductors, one of which is more readily acted upon chemically by the liquid than the other.

136. Importance of Amalgamating the Zinc. — All commercial zinc contains impurities, such as carbon, iron, etc. Figure 141 represents a zinc element having on its surface a particle of carbon a, purposely magnified. If

Fig. 141.

such a plate is immersed in dilute sulphuric acid, the particles of carbon will form with the zinc numerous voltaic circuits, and a transfer of electricity along the surface will take place. This coasting trade, as it were, between the zinc and the impurities on its surface, diverts so much from the regular battery current, and thereby weakens it. In addition to this, it occasions a great waste of chemicals, because, when the regular circuit is broken, this *local action*, as it is called, still continues. If pure zinc were used (formerly it was used), no local action would occur at any time, and there would be no consumption of chemicals, except when the circuit is closed. If mercury is rubbed over the surface of the zinc, after the latter has been dipped into acid to clean its surface, the mercury dissolves a portion of the zinc, forming with it a semi-liquid amalgam which covers up its impurities, and the amalgamated zinc then comports itself like pure zinc.

137. How Electric Energy Originates. — According to the doctrine of the conservation of energy, whenever any new form of energy is generated it is always at the expense of some other form of energy; in other words, some other form of energy is transformed into the new form. When, as in Experiment 118, you turn the plate of a Holtz machine, you feel a peculiar resistance that is not wholly due to the friction of the parts. The mechanical energy which you expend in overcoming friction is converted into heat energy. The mechanical energy which you expend in overcoming the peculiar resistance is transformed into electric energy.

We are already familiar with the fact that the chemical potential energy in a lump of coal is converted during the process of combustion into heat energy. When zinc is placed in acidulated water, a similar combustion occurs, and if a thermometer is placed in the liquid, it will show a rise of temperature as the burning progresses. If, however, the zinc is connected with the copper, or some other suitable element, there is less heat generated by the combustion, because a portion of the chemical potential energy is converted into electric energy. Electric energy originates in a voltaic cell from the conversion of chemical potential energy into this form of energy.

Section III.

BATTERIES.

138. Polarization of Elements. — If you connect any voltaic cell with a sensitive galvanoscope, such as will be described hereafter, you will find that in a few minutes after making the connection, the deflection diminishes somewhat. This is due to the collection of hydrogen at the electro-negative plate. The effect of the hydrogen is to raise the potential of this element and thereby diminish the difference of potential between the two plates. Whatever tends to diminish the difference of potential between the two elements, tends to diminish the current of electricity, and to that extent to diminish the value of a voltaic cell. This action is called, technically, *polarization* of the elements. Among the different methods that have been devised for remedying this evil, the most efficient is that in which the hydrogen is disposed of by surrounding the electro-negative element with a liquid with which the hydrogen will readily enter into combination. A good illustration of this method may be found in the action of the Bunsen battery.

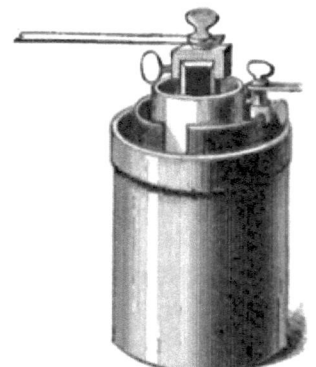

Fig. 142.

139. Bunsen Battery. — The metal employed for the electro-positive plate in this, as in nearly all batteries, is zinc. The zinc is immersed in sulphuric acid diluted with about ten times its volume of water. Inside of the hollow cylindrical zinc plate (Fig. 142) is a cup of porous unglazed earthenware. This cup contains a liquid composed of a saturated solution of potassium bichromate, or (better) sodium bichromate, mixed with one-sixth its volume of sulphuric acid. This cup serves to keep the two liquids separate, but does not prevent electrical action. In this cup is placed a bar of carbon, which is the electro-negative plate. The larger portion of the hydrogen generated by the action between the zinc and the acidulated water enters into combination with some of the constituents of the bichromate of potash, and thereby prevents in a large measure the polarization of the electro-negative element. Such a battery is called a *two-fluid* battery.

140. Grenet Battery. — In this battery a small flat plate of zinc, z (Fig. 143) is suspended between two carbon plates, CC. The carbons remain in the liquid all the time. The zinc should be drawn up out of the liquid by means of a slide rod a when the battery is not in use, as a broken circuit does not prevent the consumption of the zinc when it is in the liquid, even though the zinc is well amalgamated.

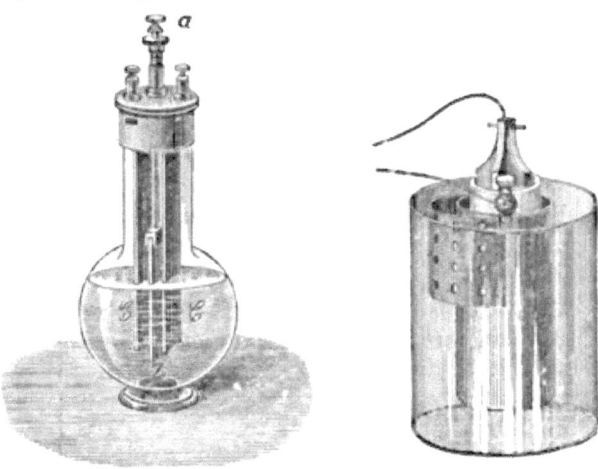

Fig. 143.   Fig. 144.

This battery gives a more energetic current for a short time than the Bunsen battery, but the carbon in this battery becomes sooner polarized, and the liquid sooner exhausted than in the Bunsen battery. It is an extremely convenient and popular battery for brief schoolroom use, as it is very energetic in its action and requires little care.

141. Daniell Battery. — One of the chief virtues of this battery (Fig. 144) is, that it polarizes less than most other kinds of batteries, and therefore gives a more constant current. The zinc is suspended in a porous cup, either in pure water, or in water to which has been added a little zinc sulphate to hasten the action when the battery is first set up. The zinc is not amalgamated. Outside the cup is a thin sheet of copper in the form of a hollow cylinder immersed in a saturated solution of copper sulphate. In a pocket near the top of the copper sheet are kept lumps of copper sulphate, which are gradually dissolved to take the place of that which is consumed by the action of the battery. This battery requires very little attention, and is largely used in England for telegraphing.

Section IV.

SOME EFFECTS PRODUCED BY AN ELECTRIC CURRENT.

142. Heating and Luminous Effects.

Experiment 120. — Connect six or eight Bunsen (or Grenet cells) *abreast* (see page 186). Attach connectors to the electrodes, and introduce between the connectors a piece of No. 30 platinum wire, less than half an inch long. The platinum wire is heated to a luminous state. Place the platinum wire over a gas burner, turn on the gas, and light it by the heat of the wire. This illustrates one of the practical uses to which the electric current is put in lighting numerous gas burners in halls and theatres. Remove the platinum wire, and introduce into the circuit an incandescent lamp of from four to six candle-power. Does this arrangement of battery render the carbon filament luminous?

Experiment 121. — Connect the eight cells *in series* (see page 187), and introduce the same lamp into the circuit. Does the filament become luminous? Remove the lamp from the circuit, and insert the platinum wire as before. Does the platinum wire become as hot as in the former arrangement of cells? Which arrangement gives the greater heating effect with the platinum wire? Which with the lamp?

143. Chemical Effects.

Experiment 122. — Take in a test-tube a quantity of an infusion of purple cabbage (the cabbage may be found at a suitable time of the year in any market) prepared by steeping its leaves until well cooked. Pour into this infusion a few drops of any alkali, such as a solution of caustic soda. The infusion is changed thereby from a purple to a green color. In another test-tube take another portion of the purple infusion. Into this pour a few drops of any acid, such as dilute sulphuric acid. The purple is changed to a red. Only acids will turn this infusion to a red and only alkalies will turn it to a green. Into a rather strong solution of sodium sulphate pour enough of the purple infusion to give it a decided color.

Pour some of this colored solution into a V-shaped glass tube (Fig. 145). Take two short pieces of copper wire covered with rubber and having strips of platinum soldered upon one of their ends for electrodes, and introduce one of these electrodes into each arm of the tube until it nearly reaches the bottom or angle of the V. By means of connectors connect the battery (of three cells in series) wires with the free extremities of these wires. The liquid between the two platinum electrodes forms a part of the circuit, so that the current of electricity passes through this portion of the liquid. Soon the liquid around the − electrode is turned green, while that around the + electrode is turned red. Evidently, decomposition of the sodium sulphate has taken place. An acid and an alkali are the results.

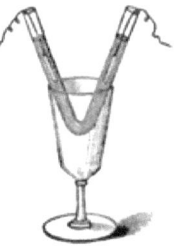

Fig. 145.

A substance that may be decomposed by electricity is called an *electrolyte*, and the process *electrolysis*. *An electrolyte must be a compound substance, and in a liquid state.* When a salt (see Chemistry, page 54) is electrolyzed, the base appears at the − pole, and the acid at the + pole.

Experiment 123.— Wet a piece of writing paper with a liquid prepared in the following manner. Dissolve by heating about three grains of pulverized potassium iodide in about a tablespoonful of water. Make a paste by boiling pulverized starch in water. Take a portion of this paste about the size of a pea, stir it into the solution. Spread the wet paper smoothly on a piece of tin, *e.g.* on the bottom of a tin basin. Press the − pole of your battery against an uncovered part of the tin. Draw the + pole over the paper. A mark is produced upon the paper as if the pole were wet with a purple ink. In this case the potassium iodide is decomposed, and the iodine combining with the starch forms a purplish blue compound.

In the experiment with the cabbage infusion you probably discovered bubbles of gas arising from this liquid, causing a foam. This is evidence that there was another decomposition going on besides that of the sodium sulphate — a sort of double decomposition. We will now take measures to collect these gases for examination.

Experiment 124. — Take a dilute solution of sulphuric acid (1 part by bulk to 20), pour some of it into the funnel (Fig. 146), so as to fill the U-shaped tube when the stoppers are removed. Place the stoppers which support the platinum electrodes tightly in the tubes. Connect with these electrodes the battery wires. Instantly bubbles of gas arise from both electrodes, accumulating in the upper part of the tube and forcing the liquid back into the tunnel. Twice as much gas arises from the − electrode as from the + electrode. Close the passage in the rubber tube by turning down the screw of the pinch-cock a. Light a splinter of fine wood, blow out the flame, leaving it glowing; remove the stopper holding the + electrode and introduce the glowing splinter into the gas in this arm of the tube. It relights and burns vigorously, showing that the gas is oxygen. (See Chemistry, page 19.) Platinum electrodes are used, otherwise a portion of the oxygen carried to the + electrode would not be set free, but would oxydize the metal (e.g. copper), instead of appearing as a gas in this arm of the tube. Fill this arm of the tube with water and stopper it. Invert the U-tube; the gas in the other arm of the U-tube collects in the bend of the tube and in the small branch tube. Light a match, remove the

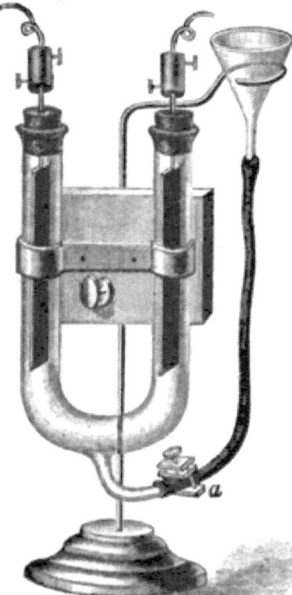

Fig. 146.

rubber tube, and quickly hold the match at the orifice of the branch tube. The gas burns. (See Chemistry, page 25.) It is *hydrogen*. This operation is commonly called "decomposing water by electricity." See if you can "decompose water" with your battery of three cells connected *abreast*.

Experiment 125. — This delightful experiment may be performed by the teacher, or an experienced pupil, before the class. Take about one-fourth of a teacupful of water, dissolve in it about two grams of silver nitrate. Do not wet the hands with the solution, as it will stain them black. Nearly fill the electrolysis tank (Fig. 147) which accompanies the porte-lumière (page 310).

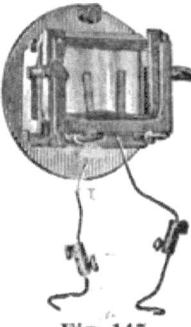
Fig. 147.

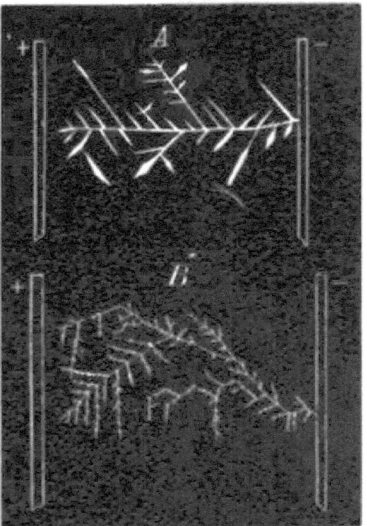
Fig. 148.

Arrange a battery of two cells in series. Place the tank in position in the porte-lumière to project it on a screen in a dark room. Connect the battery wires with the electrodes in the tank. A beautiful deposit of silver will be made on the − electrode, spreading therefrom toward the + electrode, and bearing a strong resemblance to vegetable growth; hence it is called the "silver tree." In Figure 148, A represents a silver tree deposited from a weak solution and B from an extremely weak solution.

144. Physiological Effects.

Experiment 126. — Take a single Bunsen cell and place its electrodes each side of the tip of the tongue. A slight stinging (not painful) sensation is felt, followed by a peculiar acrid taste.

When a battery is known not to be very powerful, the tongue serves as a convenient galvanoscope to determine whether the battery is in working condition.

145. Magnetic Effects.

Experiment 127. — Take the iron rod having an insulated wire wound around it, and connect the extremities of the wire with the battery wires; in other words, introduce this wire into the circuit. Bring a nail (Fig. 149) or other piece of iron near one end of the rod. The rod attracts the nail with considerable force, and this nail will attract other nails. The rod has all the properties of a magnet, as will be seen hereafter. Break the circuit. The iron rod instantly loses its magnetic force, and the nails drop.

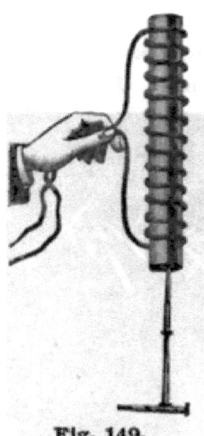

Fig. 149.

The iron rod is called a *core*, the coil of wire a *helix*, and both together an *electro-magnet*. In order to take advantage of the attraction of both ends or *poles* of the magnet, the rod is most frequently bent into a U-shape (A, Fig. 150), and then it is called a horse-shoe magnet. More frequently two iron rods are used, connected by a rectangular piece of iron, as a in B of Figure 150. The method of winding is such that if the iron core of the horse-shoe were straightened, or the two spools were placed together end to end,

 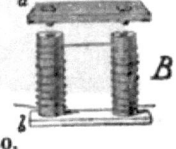

Fig. 150.

one would appear as a continuation of the other. A piece of soft iron, b, placed across the ends and attracted by them, is called an *armature*. The piece of iron, a, is called a *yoke*.

Experiment 128. — Arrange a battery of four cells *in series*. Introduce into the circuit an electro-magnet wound with a long, fine wire (having a resistance of not less than 25 ohms.[1] Ascertain approximately the force required to pull an armature (*e.g.* a large nail) off the poles.

Next remove this electro-magnet, and introduce into the circuit in its place an electro-magnet wound with coarse wire (which has a resistance not exceeding 1 ohm). See, by pulling, with what force it holds a nail on one of its poles.

Experiment 129. — Arrange a battery of four cells *abreast*. Introduce into the circuit the fine wire magnet. See with what force it

---

[1] See page 178.

holds its armature. Which arrangement of cells produces the greater magnetic power with this electro-magnet?

Next introduce in its place the low-resistance electro-magnet and find with what force it holds the nail. Which arrangement of cells produces the greater magnetic power in this magnet?

Important Lesson: The results of our experiments thus far teach, that *the arrangement of a battery of several cells and the apparatus used should be adapted to each other.*

It is apparent that, if there are rules or laws which will enable a person who would use an electric current for experimental or industrial purposes, to determine by calculation just what is the best method of arrangement in any given case, it is of the utmost importance that these laws should be understood.

Section V.

ELECTRICAL MEASUREMENTS.

The wonderful developments which have been made in recent years in electrical science, and which have led to the employment of electric energy in connection with a great diversity of industrial arts, are almost wholly due to a better understanding of what electrical measurements can be made, and how to make them. Indeed, little of a *practical nature* can be done without some acquaintance with the methods of making these measurements.

146. Some Technical Terms. — A quantity of water may be measured either in quarts or pounds; *i.e.* by its volume or weight. Although electricity has neither volume nor weight, yet it has that which answers strictly to the term *quantity*, and the quantity can be measured by suitable means. The unit employed for the measurement of a quantity of electricity is called a *coulomb*. A stream of water flowing through a pipe might be described by stating the number of quarts which flow through the pipe, or past any point in the pipe, in a minute. In a similar manner, we describe an electric current by stating the number of coulombs that pass through a conductor, or that pass a given point of a conductor, in a second.

The quantity of electricity passing through a conductor in a given time, in other words *the rate of flow*, determines *the strength of the current*. When the quantity passing is *one coulomb*[1] *per second*, the strength of the current is said to be one *ampère*. A current of 10 coulombs per second has a strength of 10 ampères. The ampère is the unit for measuring current strength. There is no unit analogous to this for measuring liquid currents. It should be observed that the term strength refers only to the *quantity of electricity passing*, and not to the energy of the current.

As we might calculate the energy of a current of water by multiplying the weight of water falling by the distance it falls, so if we represent by C the strength of current in ampères, and by V the electro-motive force or difference of potential in volts (see next paragraph), then

$$CV = \text{energy of current,}$$

which is expressed in a unit called an *ampère-volt* (or *watt*), much as we express mechanical energy in foot-pounds. This amount of energy per second is equivalent to about $\frac{1}{746}$ horse-power.

From this formula we infer that when the electro-motive force remains the same, the energy of a current varies as its strength; and when the current strength does not change, the energy varies as the electro-motive force. As indicated above, difference of potential and electro-motive force are measured in a unit called a *volt*. For our pur-

---

[1] *A coulomb* is the quantity of electricity delivered by a one-ampère current in one second.

pose it will answer to consider a volt as the electro-motive force of a Daniell's cell; *i.e.* it is about the difference of potential between the zinc and the copper of this cell.

Section VI.

C.G.S. MAGNETIC AND ELECTRO-MAGNETIC UNITS.

[This section is intended to assist the student who is ambitious to read technical works on electricity, but, like other matter in fine print, it is not included in the course of study prescribed in this book.]

147. Magnetic Units. — These units are based on the forces exerted between two magnetic poles. They form the basis for the electrical units adopted by the Congress of Electricians, held at Paris in 1871.

Unit Magnetic Pole. — A unit magnetic pole is one which repels a similar pole placed at a distance of 1^{cm} with a force of 1 dyne. It has no special name; its dimensions are $M^{\frac{1}{2}}L^{\frac{3}{2}}T^{-1}$.

Unit of Intensity of a Magnetic Field. — The intensity of a magnetic field is one C.G.S. unit when the force which acts on a unit magnetic pole in this field is 1 dyne. Its dimensions are $M^{-\frac{1}{2}}L^{\frac{1}{2}}T^{-1}$.

148. Electro-Magnetic Units and Practical Units. —
Unit of Current Strength: — A current has the strength of one C.G.S. unit, if, in passing through a circuit 1^{cm} long, bent into the form of an arc of a circle of 1^{cm} radius (so as to be always 1^{cm} away from the magnet-pole), it exerts a force of 1 dyne on a unit magnet-pole placed at the center.

Unit of Quantity: — The quantity of electricity which passes through a circuit in one second when the strength of the current is one C.G.S. unit.

Unit of Electro-motive Force: — The E.M.F. necessary in order that a unit of quantity may do the work of an erg. [$W = QE$.]

Unit of Resistance: — A conductor has a resistance of one C.G.S. unit when a unit difference of potential between its two ends causes a unit of current to pass through it.

Inasmuch as in practice the employment of these units leads to the use of very large numbers, units have been adopted which are decimal multiples of the C.G.S. units. They have received special names and are known as the *practical units*.

TABLE OF ELECTRO-MAGNETIC UNITS.

QUANTITIES.	SYMBOL.	NAME OF PRACTICAL UNITS.	NO. OF C.G.S. UNITS IN ONE PRACTICAL UNIT.	DIMENSIONS OF UNIT.
Resistance	R	Ohm	10^9	LT^{-1}
Electro-motive force	E	Volt	10^8	$M^{\frac{1}{2}}L^{\frac{3}{2}}T^{-2}$
Current Strength .	C	Ampère	10^{-1}	$M^{\frac{1}{2}}L^{\frac{1}{2}}T^{-1}$
Quantity	Q	Coulomb	10^{-1}	$M^{\frac{1}{2}}L^{\frac{1}{2}}$

Section VII.

GALVANOMETERS.

149. Introductory Experiments.

Experiment 130. — Wind a battery wire lengthwise once around a book, and place the book either above or below and near to a magnetic needle, and hold the book in such a position that that portion of the current which circulates around the book will have a northerly and southerly direction. Notice the extent of the deflection of the needle. Then wind the wire closely 20 or 30 times around the book, and hold it in the same position, and at the same distance from the needle as before. The needle, now that the current is carried several times past it, makes a larger deflection; consequently the effect of several windings is to render the needle more sensitive to weak currents.

Experiment 131. — Connect two cells abreast, and once more hold the book with its many turns of wire near the needle, as in the last experiment. The deflection is larger than before, which is due to the fact that the two cells give a stronger current than one cell.

It thus seems that a galvanoscope, in addition to its other uses, may indicate the strength of a current, and when properly constructed to measure the relative strength of currents it is called a *galvanometer*.

150. Tangent Galvanometer. — The galvanometer G, represented in Figure 151, has a magnetic needle about ⅓ inch long and an indicator of light aluminum wire about 3.5 inches long, resting upon and parallel with the needle. The whole is suspended from a brass frame by a very fine, untwisted, silk fiber, just over a coil of wire such as was formed by winding a wire about the book. Between the needle and coil is a card containing a circle divided into halves by a diameter parallel with the wires of the coil below. Each extremity of this diameter is numbered zero. Each semicircle is divided into halves, and each quarter circle is divided into ninety degrees and numbered each way from zero to the ninetieth degree. The whole is covered with a glass case to prevent disturbance by currents of air.

When the needle of a galvanometer is short in comparison with the length of its coil *the strength of currents varies as the tangents of the angles of deflection.* Such a galvanometer is called a *tangent galvanometer.* For example, suppose that the deflections produced in *the same* tangent galvanometer by two currents are 80° and 70°. Consulting the Table of Tangents in Appendix, C, we find the tangents of these angles are respectively 5.67 and 2.75; hence the former current is $(5.67 \div 2.75 =)\ 2 +$ times as strong as the latter.

<small>The student should understand that the galvanometer described above, is not, strictly speaking, a standard tangent galvanometer. The manifold uses to which galvanometers are put in a physical laboratory, properly require a variety of instruments, which would make an equipment very expensive. The galvanometer here described answers very well all the purposes of this book. The results obtained by its use are approximately those which would be obtained by a standard tangent galvanometer of the usual form, in which the needle is suspended at the center of a large circular coil of wire.</small>

151. Galvanometer with an Astatic Needle. — This needle is much more sensitive to weak currents than the needle described above.

It consists of two magnetic needles fastened to a common axis, but having their poles reversed, so that, for example, the + pole of one is over the — pole of the other. It is suspended by a silk fiber, so that one of the needles may rotate within the coil while the other rotates above the coil.

The current acts upon both needles to turn them in the same direction. Moreover the current both above and below acts in the same direction on the needle which is suspended within the coil, hence the astatic needle is much more sensitive than a single needle. The needle does not point north and south like the ordinary needle, but more nearly east and west.

Section VIII.

RESISTANCE OF CONDUCTORS.

152. External Resistance.

Experiment 132. — Introduce into a circuit a galvanometer, and note the number of degrees the needle is deflected. Then introduce into the same circuit the wire on the spool numbered 4 on the platform, S (Fig. 151). (The wire on any one of the five spools on this platform can at any time be introduced into a circuit, by connecting the battery wires with the binding screws on each side of the spool to be introduced.)

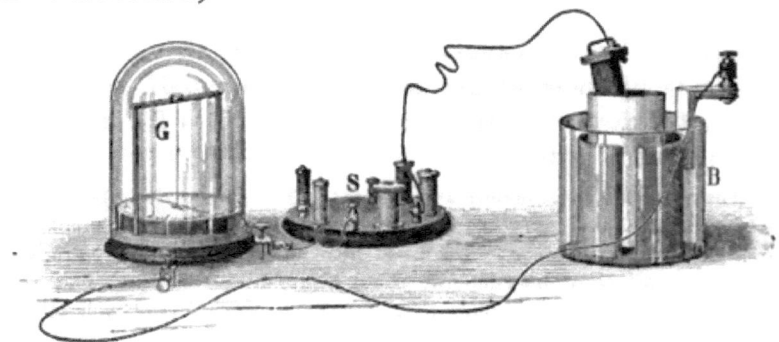

Fig. 151.

The deflection is now less than before. The copper wire on this spool is 16 yards in length; its size is No. 30 of the Brown and Sharpe wire gauge. When this spool is in circuit, the circuit is 16 yards longer than when the spool is out. The effect of lengthening the circuit is to weaken the current, as shown by the diminished deflection.

Experiment 133. — Next, substitute Spool 2 for Spool 4. This contains 32 yards of the same kind of wire as that on Spool 4. The deflection is still smaller.

The weakening of the current by introducing these wires is caused by the resistance which the wires offer to the current, much as the friction between water and the interior of a pipe impedes, to some extent, the flow of water through it. The longer the pipe the greater is the resistance to the flow.

If the wire on the spools had been the only resistance in the circuit, then, when Spool 2 was in the circuit, the resistance of the circuit would have been double the resistance that it was when Spool 4 was in the circuit, and the current, with double the resistance, would have been half as strong.

(1) *Other things being equal, the resistance of a conductor varies as its length.*

Experiment 134. — Next substitute Spool 1 for Spool 2. This spool contains 32 yards of No. 23 copper wire, — a thicker wire than that on Spool 2, but the length of the wire is the same. The deflection is now greater than it was when Spool 2 was in circuit. This indicates that the larger wire offers less resistance.

Careful experiments show that (2) *the resistance of all conductors varies inversely as the areas of their cross sections. If the conductors are cylindrical it varies inversely as the square of their diameters.*

Experiment 135. — Substitute Spool 5 for Spool 1, and compare the deflection with that obtained when Spool 4 was in the circuit. The deflection is smaller than when Spool 4 was in circuit. The wire on these two spools is of the same length and size, but the wire of Spool 5 is German silver. It thus appears that German silver offers more resistance than copper.

(3) *In estimating the resistance of a conductor, the specific resistance of the substance must enter into the calculation.* (See Table of Specific Resistances, Appendix, D.)

The resistance of metal conductors increases slowly with the temperature of the conductor. The resistance of German silver is affected less by changes of temperature than that of most metals; hence its general use in standards of resistance.

153. Internal Resistance.

Experiment 136. — Connect the copper and zinc strips used in Experiment 114 with the galvanometer, and introduce the strips into a tumbler nearly full of acidulated water. Note the deflection. Then raise the strips, keeping them the same distance apart, so that less and less of the strips will be submerged. As the strips are raised, the deflection becomes smaller. This is caused by the increase of resistance in the liquid part of the circuit, as the body of liquid lying between the two strips becomes smaller. The resistance of the liquid part of the battery is called *internal resistance*, in distinction from that of the rest of the circuit, which may be regarded as *external resistance*.

(4) *The internal resistance of a circuit varies inversely as the area of the cross section of the liquid between the two elements.*

In a large cell the area of the cross section of the liquid between the elements is larger than in a small cell, consequently the internal resistance is less. This is the only way in which the size of a cell affects the current.

154. Measurement of Resistance; The Ohm.
— Resistance is measured by a unit called an *ohm*. An ohm is the resistance of about 9 inches of No. 30 (B. & S. G.) German silver wire, or about 9.3 feet of No. 30 copper wire at ordinary temperature.

155. Description of the Rheostat.
— Figure 152 represents a wooden box containing what is equivalent to a series of coils of German silver wire, whose resistance ranges from .01 ohm to 100 ohms. Each of these coils is connected with a brass stud on the top of the box.

Three switches, A, B, and C, so connect the coils with the binding screws *a* and *b* that a current can be sent through any three coils at the same time by moving the switches on to the proper studs. The resistance in ohms of each coil is marked on the box near its stud. When the three switches rest upon studs marked 0, the current meets with no appreciable resistance in passing through the box, but any desired resistance within the range of the instrument can be introduced by moving the switches on to the studs, the sum of whose resistances is the resistance required. This instrument is called a *rheostat*.

Fig. 152.

Experiment 137. — Measure in ohms the resistance of the wire on each one of the spools used above, as follows : — Introduce into circuit (as in Figure 151) a galvanometer and the spool whose resistance is sought. Note the deflection in degrees. Then remove the spool, and introduce the rheostat in its place. Place all the switches on the zero studs. The deflection of the galvanometer needle is now evidently greater than when the spool was in circuit. Move the switches, throwing in or taking out resistance (much as you use weights in weighing), until the deflection becomes the same as the deflection was when the spool was in circuit. It is evident that the sum of the resistances, as indicated by the three switches, must be the same as the required resistance of the wire on the spool.

In the same manner, measure the resistance of the electro-magnets of telegraph sounders, relays, incandescent lamps, etc.

The method of measuring resistance given above is called the *method by substitution or balancing*. The results obtained by this method are accurate only on condition that the electro-motive force and internal resistance of the battery remain sensibly constant throughout the operation. This rarely happens, so that the results obtained can be regarded as only *approximately* correct. When great accuracy is required, it is necessary that some means of measuring should be adopted in which the fluctuations of the battery will not affect the results. This difficulty is obviated by the use of the invaluable instrument called (from the name of its inventor) the Wheatstone bridge.

156. Wheatstone Bridge.—Figure 153 represents a perspective view of the bridge (as modified by the Author), and Figure 154 represents a diagram of the essential electrical connections. The battery wires are connected with the bridge at the binding screws, BB'. A galvanometer g is connected at GG', a rheostat r at RR, and the object x, whose resistance is sought, at XX.

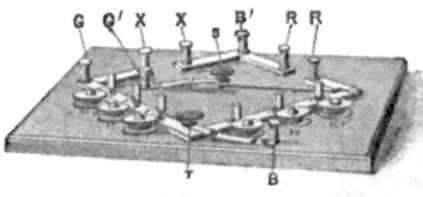

Fig. 153.

On closing the circuit by pressing on the knob T the current, we will suppose, enters at B; on reaching the point A it divides, one part flowing *via* the branch AGB', and the other *via* the branch ADB'. If points D and G in the two branches have different potentials and a connection is made between them through the galvanometer, g, by pressing on the knob S, there will be a current through this bridge wire and through the galvanometer, and a deflection of the needle will be produced. But if the points D and G have the same potential, there will be no cross current through the bridge wire and no deflection. Now it can be demonstrated that points D and G will have the same potential when R (the resistance) of AD : R of DB' :: R of AG : R (the unknown resistance) of GB'. Between A and D and A and G there are three coils of wire having resistances respectively of 1, 10, and 100 ohms. One or more of these coils are introduced into the circuit by removing the corresponding plugs a, b, c, d, e, and f. As the other connections between A and D, and A and G, have no appreciable resistance, being for the most part short brass bars, the only practical resistance between these points is that intro-

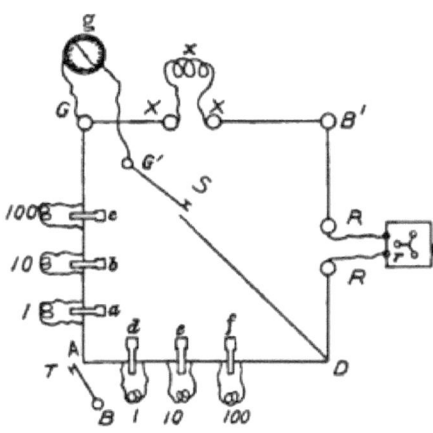

Fig. 154.

duced at will through the coils. Similarly between points D and B′, the only practical resistance is that introduced at will through the rheostat, and between points G and B′ the resistance is the resistance (r) sought.

It is apparent, then, that in using the bridge after the connections are properly made through the several instruments and certain known resistances are introduced between A and D, and A and G, we have simply to regulate the resistance through the rheostat so that there will be no deflection in the galvanometer; then we are sure that the above proportion is true. The first three terms of the proportion being known, the fourth term, which is the resistance sought, is computable.

In using the instrument, observe the following directions. (1) Always close the circuit at T before closing the bridge connections at S. (2) Introduce between A and D, and A and G, resistance as nearly equal to the resistance (r) sought as practicable, as the galvanometer is then most sensitive. If you have no conception what the unknown resistance is, it is best to begin by using high resistances. (3) The sensitiveness of the galvanometer may be greatly increased by placing on the table a bar magnet in the magnetic meridian with its north-seeking pole turned toward the north-seeking pole of the needle.

Experiment 138. — Measure the resistances of the several spools of wire used above, — electro-magnets, electric lamps, etc., — using the bridge. Place the switches of the rheostat on the zero studs. Make connections as in the description above. Then close the circuit at T, and afterward the bridge at S. There will probably be a deflection in the galvanometer. Regulate the resistance through the rheostat, throwing in or taking out resistance according as one or the other tends to reduce the deflection (the process is much as in weighing), until there is no deflection. Then compute the resistance sought according to the above proportion. Compare the results with those obtained by the process of substitution.

Experiment 139. — Measure the resistance of the human body. Let some person grasp in his dry hands two metallic handles, such as are used in giving shocks; connect the handles by wires at X X. Introduce 100 ohms between A and G, and 1 or 10 ohms between A and D, and proceed as hitherto.

The cuticle, or dry outer skin of the body, offers great resistance. Let the same person wet his hands, and measure the resistance again, and ascertain how much the wetting of the cuticle reduces the resistance. Then let the person wet his hands with strong salt brine, and once more measure the resistance.

Section IX.

ELECTRO-MOTIVE FORCE OF DIFFERENT BATTERIES;
OHM'S LAW.

157. Electro-Motive Force of Different Batteries. — If a galvanometer is introduced into a circuit with different battery cells, *e.g.* Bunsen, Grenet, Daniell, etc., very different deflections will be obtained, showing that the different cells yield currents of different strength. This may be due in some measure to a difference in their internal resistance, but it is chiefly due to the difference in their electro-motive force. We learned (page 161) that difference of electro-motive force is due to the difference of the chemical action on the two plates used, and this depends largely upon the nature of the substances used. It is wholly independent of the size of the plates; hence the electro-motive force of a large battery cell is no greater than that of a small one of the same kind. Consequently any difference in strength of current yielded by battery cells of the same kind, but of different sizes, is due wholly to a difference in their internal resistance.

The electro-motive force of the Bunsen, Grenet, and Daniell cells are respectively about 1.8, 2, and 1 volts.

In consequence of polarization of the plates, the electro-motive force of most batteries diminishes more or less rapidly after beginning to work. For example, the current of the Leclanché battery weakens so rapidly that it can be used only in cases in which the battery is required to work only for a few minutes at a time, such as for ringing annunciator bells, telephony, etc.

158. Ohm's Law. — *The strength of current in any voltaic circuit varies directly as the electro-motive force and in-*

versely as the total resistance of the circuit. Likewise, the current between any two points varies as the difference of potential between those points, and inversely as the resistance to be overcome. This law is usually expressed in the form of the mathematical formula

$$C = \frac{E}{R}; \text{ whence } E = RC, \text{ and } R = \frac{E}{C},$$

in which C represents the strength of current, E the electro-motive force, and R the entire resistance. The above fraction $\frac{E}{R}$, when the external resistance is considered separately from the internal, must be converted thus; calling the former R, and the latter r, the expression becomes

$$C = \frac{E}{R + r}.$$

If a cell has $E = 1$ volt, and $r = 1$ ohm, and the connecting wire is short and stout, so that R may be disregarded, then the current has a value of one ampère. In other words, an ampère might be defined as the strength of current which an electro-motive force of one volt will maintain through a resistance of one ohm.

EXERCISES.

1. What E.M.F. is required to maintain a current of one ampère through a resistance of one ohm?

2. An E.M.F. of 10 volts will maintain a current of 5 ampères through what resistance?

3. What current ought an E.M.F. of 20 volts to maintain through a resistance of 5 ohms?

4. A volt-meter applied each side of an electric lamp shows a difference of potential of 40 volts; what current flows through the lamp, if it has a resistance of 10 ohms?

5. The resistance between two points in a circuit is 10 ohms. An

ammeter (an instrument which measures the strength of a current in ampères) shows that there is a current strength in the circuit of 0.5 ampère; what is the difference in potential between the points?

6. What current will a Bunsen cell furnish when $r = 0.9$ ohm (about the resistance of a quart cell), $E = 1.8$ volts, and $R = 0.01$ ohm (about the resistance of 3 ft. of No. 16 wire)?

Section X.

DIVIDED CIRCUITS; METHODS OF COMBINING VOLTAIC CELLS.

159. Divided Circuits; Shunts.

Experiment 140. — Make a divided circuit as in Figure 155 (using double connectors a and b). Insert a galvanometer, G, in one branch and a rheostat, R, in the other. The current, when it reaches a, divides, a portion traversing one branch through the galvanometer, and the remainder passes through the other branch and the rheostat. Either branch may be called a *shunt* to the other. Increase gradually the resistance in the rheostat. The result is that it throws more of the current through the galvanometer, as shown by the increase of deflection.

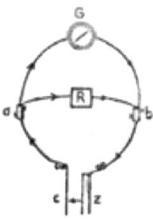

Fig. 155.

In a divided circuit the current divides between the paths inversely as their resistances. For example, if the resistance of the rheostat above is 4 ohms and the resistance in the galvanometer is 1 ohm, then four-fifths of the current will traverse the latter and one-fifth the former.

Suppose that the rheostat and galvanometer are removed from the shunts, and that the shunts are of the same length, size, and kind of wire, and consequently have equal resistances. Using the two wires instead of one to connect a

and b is equivalent to doubling the size of this portion of the conductor; consequently the resistance of this portion is reduced one-half.

Generally, *the joint resistance of two branches of a circuit is the product of their respective resistances divided by their sum.* (For demonstration of this law, see Gray's Absolute Measurements in Electricity, page 84.)

160. Methods of Combining Cells.

Experiment 141. — Take two Bunsen cells, and connect the two zinc plates by a wire. Then connect each of the carbon plates with a galvanometer. The current from the two cells, if there were any, would flow in opposite directions. But you find that there is either no deflection in the galvanometer, or at most a very small one, and this shows either that there is no current or that the current is very weak. The reason is evident. You have connected two carbons, which have theoretically the same potential, through the galvanometer; consequently there should be no current between them. The cells are said to be *connected in opposition*.

A very simple way of showing that a large cell has no greater electro-motive force than a small one is to connect two such cells in opposition through a galvanometer, or, what answers the same purpose, raise the zinc of one of two cells of the same size, connected in opposition, nearly out of the liquid. The absence of a current shows that the two carbons have the same potential, and consequently their electro-motive force is the same.

A number of cells connected in such a manner that the currents generated by all have the same direction constitutes *a voltaic battery*.

The object of combining cells is to get a stronger current than one cell will afford. We learn from Ohm's law that there are two, and only two, ways of increasing the strength of a current. It must be done either by increasing the E.M.F. or by decreasing the resistance. So we combine cells into batteries, either to secure greater E.M.F., or to diminish the internal resistance. Unfortunately, both purposes cannot be accomplished by the same method.

161. Batteries of Low Internal Resistance. — Figure 156 represents three cells having all the carbon (c) plates electrically connected with one another, and all the zinc (z) plates connected with one another, and the triplet carbons are connected by the leading-out wires through a galvanometer with the triplet zincs.

It is easy to see that through the battery the circuit is divided into three parts, and consequently the conductivity in this part of the circuit, according to the principle stated in § 159, must be increased threefold; in other words, the internal resistance of the three cells is one-third of that of a single cell. This is called connecting cells "abreast,"

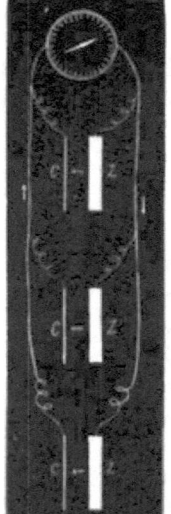

Fig. 156.

or "in multiple arc," and the battery is called a "battery of low internal resistance." The resistance of the battery is decreased as many times as there are cells connected in "arc," but the E.M.F. is that of one cell only.

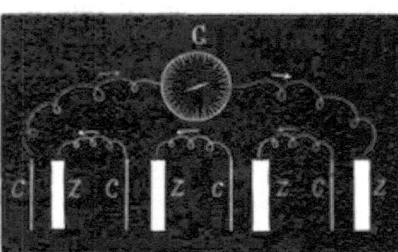

Fig. 157.

162. Batteries of High Internal Resistance and Great E.M.F. — Figure 157 represents four cells having the carbon or $+$ plate of one connected with the zinc or

— plate of the next, and the + plate at one end of the series connected by leading-out wires through a galvanometer with the — plate at the other end of the series. It is evident that the current in this series traverses the liquid four times, which is equivalent to lengthening the liquid conductor four times, and, of course, increasing the internal resistance fourfold. But, while the internal resistance is increased, *the E.M.F. of the battery is increased as many times as there are cells in series.* The gain by increasing the E.M.F. more than offsets, in many cases (always when the internal resistance is a small part of the whole resistance of the circuit), the loss occasioned by increased resistance.

163. Best Arrangement of Cells.

Experiment 142. — Introduce into circuit with a single Bunsen cell a rheostat and a galvanometer. Throw a resistance of (say) 50 ohms into the circuit by means of the rheostat. Note the deflection. Then add another cell, in series, to the cell already in use. The deflection is considerably increased. Other cells may be added with similar results.

Experiment 143. — Connect the two cells abreast, keeping the same resistance in the rheostat. The deflection is only a very little greater than that caused by a single cell.

Experiment 144. — Connect a single cell with a galvanometer[1] of low resistance, so that the whole external resistance may be less than the resistance of the single cell. Note the deflection. Then introduce another cell abreast. The deflection is considerably increased.

Experiment 145. — Connect the same cells in series. The deflection differs but little from that produced by a single cell.

Hence (1) *when the external resistance is large, connect cells in series;* (2) *when the external is less than the internal resistance, connect cells in arc.*

---

[1] The galvanometers furnished by the author have a resistance of about one ohm. The internal resistance of a Bunsen cell can easily be made greater than this if the cell is filled not more than one-fifth full with liquid.

The maximum current with a given number of cells through a given external resistance is attained when the external and internal resistances are most nearly equal.

Caution: — Never increase the external resistance for the purpose of making the two resistances equal.

EXERCISES.

In the following exercises, whenever a Bunsen cell is mentioned it may be understood to be a quart cell, having a resistance of about 0.9 ohm. Its E.M.F. is about 1.8 volt.

1. (a) When is a large cell considerably better than a small one? (b) When does the size of the cell make little difference in the current?

2. If you have a dozen quart cells, how can you make them equivalent to one 3 gallon cell?

3. If a battery of 10 cells has an E.M.F. ten times greater than that of a single cell, why will not the battery yield a current ten times as strong?

4. (a) The internal resistance of ten cells, connected in arc, is what part of that of a single cell? (b) If the cells were connected, in series, how would the resistance of the battery compare with that of one of its cells? (c) How would the E.M.F. of the latter battery compare with that of a single cell?

5. What current will a single Bunsen cell furnish through an external resistance of 10 ohms?

6. What current will 8 Bunsen cells, in series, furnish through the same resistance?

SOLUTION: $\dfrac{E}{R+r} = \dfrac{1.8 \times 8}{10 + 0.9 \times 8} = 0.83 +$ ampère.

7. What current will 8 Bunsen cells, in arc, furnish through the same external resistance?

SOLUTION: $\dfrac{E}{R+r} = \dfrac{1.8}{10 + (0.9 \div 8)} = 0.17 +$ ampère.

8. What current will a Bunsen cell furnish through an external resistance of 0.4 ohm?

9. What current will a battery of two Bunsen cells, in series, furnish through the same resistance as the last?

10. What current will two cells, in arc, furnish through the same resistance?

Section XI.

TRANSFORMATION OF ELECTRIC ENERGY INTO HEAT.

164. Transformation Inside and Outside a Battery.
Experiment 146. — Arrange two batteries, each consisting of two (Bunsen) cells connected in arc. Use thick copper wire for leading-out wires. Attach, by means of a connector, a piece of platinum wire about 1 inch long to one of the electrodes of one of the batteries. Place a thermometer in the dilute acid of one cell of each of the batteries. Close the circuits of both batteries (one through the platinum wire) at the same moment. Watch for changes of temperature in the liquids. The temperature of the battery which is not in circuit with the platinum wire rises faster than the other.

That portion of the energy of an electric current which is not transformed into heat, or other kind of work, in other parts of the circuit, is transformed into heat in the battery.

The transformation is greatest where the resistance is greatest. The platinum wire being small, and having a relatively large specific resistance, offers much more resistance to the current than the copper wire, consequently it becomes much hotter. Much of the electric energy being transformed into heat in the platinum wire, there is less to be transformed in the battery; consequently the battery remains comparatively cool.

Section XII.

MAGNETS AND MAGNETISM.

165. Law of Magnets. — Suspend by fine threads in a horizontal position two stout darning-needles which have been drawn in the same direction (*e.g.* from eye to point) several times over the same pole (better the — pole) of a powerful electro-magnet. These needles, separated a few feet from each other, take positions parallel with each other, and both lie in a northerly and southerly direction with the points of each turned in the same direction.

That point in the Arctic zone of the earth toward which magnetic needles point is called the north magnetic pole of the earth. That end of a needle which points toward the north magnetic pole of the earth is called the *north-seeking, marked, or + pole* (inasmuch as this is the end that is always marked for the purpose of distinguishing one from the other). That end of the needle which points southward is called the *south-seeking, unmarked, or — pole.*

Experiment 147. — Bring both points near each other; they repel each other. Bring both eyes near each other; they likewise repel each other. Bring a point and an eye near each other; they attract each other.

Like poles of magnets repel, unlike poles attract one another.

166. Magnetic Transparency and Induction.

Experiment 148. — Interpose a piece of glass, paper, or wood-shaving between the two magnets. These substances are not themselves perceptibly affected by the magnets, nor do they in the least affect the attraction or repulsion between the two magnets.

Substances that are not susceptible to magnetism are said to be *magnetically transparent*. When a magnet causes another body, in contact with it or in its neighborhood, to become a magnet, it is said to *induce* magnetism in that body; *i.e.* it *influences it to be like itself*. As attraction, and never repulsion, occurs between a magnet and an unmagnetized piece of iron or steel, it must be that the magnetism induced in the latter is such that opposite poles are adjacent; that is, a N or + pole induces a S or − pole next itself, as shown in Figure 158.

Fig. 158.

167. Polarity.

Experiment 149. — Strew iron filings on a flat surface, and lay a bar-magnet on them. On raising the magnet, it is found that large tufts of filings cling to the poles, as in Figure 159, especially to the edges; but the tufts diminish regularly in size from either pole towards the centre, where none are found.

Magnetic attraction is greatest at the poles, and diminishes towards the center, where it is nothing, or the center of the bar is neutral. The dual character of the magnet, as exhibited in its opposite extremities, is called *polarity*, and magnetism is styled a *polar force*. If a magnet is broken, each piece becomes a magnet with two poles and a neutral line of its own.

Fig. 159.

168. Coercive Force.

— It is more difficult to magnetize steel than iron; on the other hand, it is difficult to demagnetize steel, while soft iron loses nearly all its magnetism as soon as it is removed from the influence of the inducing body. The quality of steel by which it at first resists the power of magnets, and resists the escape of magnetism which it has once acquired, is called *coercive force*. *The harder steel is, the greater is its coercive force.* Hence, highly tempered steel is used for *permanent* magnets. Hardened iron possesses considerable coercive force;

hence, the cores of electro-magnets should be made of the *softest* iron, that they may acquire and part with magnetism instantaneously.

169. Forms of Artificial Magnets. — Artificial magnets, including permanent magnets and electro-magnets, are usually made in the shape either of a straight *bar*, or of the letter U, called the *horseshoe*, according to the use made of them. If we wish, as in the experiments already described, to use but a single pole, it is desirable to have the other as far away as possible; then, obviously, the bar magnet is most convenient. But if the magnet is to be used for lifting or holding weights, the horseshoe form is far better, because the attraction of both poles is conveniently available, and because their combined power is more than twice that of a single pole. Magnets, when not in use, ought always to be protected by armatures (A, Fig. 160) of soft iron; for, notwithstanding the coercive power of steel, they slowly part with their magnetism. But when an armature is used, the opposite poles of the mag-

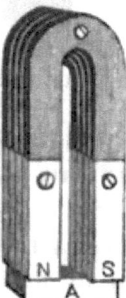

Fig. 160.

net and armature being in contact with one another, *i.e.* N with S, they serve to bind one another's magnetism. Thin bars of steel can be more thoroughly magnetized than thick ones. Hence, if several thin bars (Fig. 160) are laid side by side, with their corresponding poles turned in the same direction and then screwed together, a very powerful magnet is the result. This is called a *compound magnet*.

170. Attraction and Repulsion between Currents: Laws of Currents.

Experiment 150. — Figure 161 represents a portion of a divided circuit. The lower ends of the wires dip at the lower extremities one-sixteenth of an inch into mercury, and they are so suspended that they are free to move toward or from each other. Send a current of a battery of two or three Bunsen cells, in arc, through this divided circuit. The two portions of the current travel in the same direction and parallel with each other, and the two wires at the lower extremities move toward each other, showing an attraction.

Experiment 151. — Make the connections (Fig. 162) so that the current will go down one wire and up the other. They repel each other.

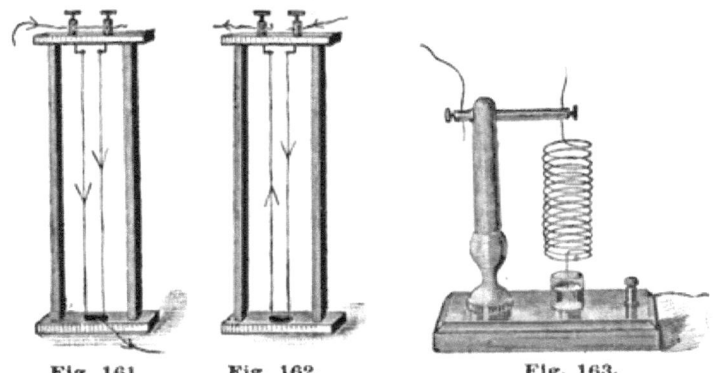

Fig. 161.   Fig. 162.   Fig. 163.

Experiment 152. — Send a current through the spiral wire represented in Figure 163. Here the current flows nearly parallel with itself, and the attraction causes the coil to contract and to be lifted out of the cup of mercury below. But the instant it leaves the mercury the circuit is broken, the current and attraction cease, and the wire dips into the mercury again. Thus rapid vibratory motion of the coil is produced.

First Law of Currents. — *Parallel currents in the same direction attract one another; parallel currents in opposite directions repel one another.*

Experiment 153. — Figure 164 represents a small battery floating on water. The wire of the battery is wound into a horizontal coil. In a few minutes after the battery is floated it will take a position so that its coil will point north and south, like a magnetic needle.

Place the wire of another battery over and parallel with the coil, so that the two currents will flow in planes at right angles with each other. The coil is deflected like a magnetic needle. A careful examination will disclose the fact that not only have the planes in which the two currents flow become parallel, but that the current in the half of the coil (where the influence due to proximity is greatest) *flows in the same direction that the current above it flows.*

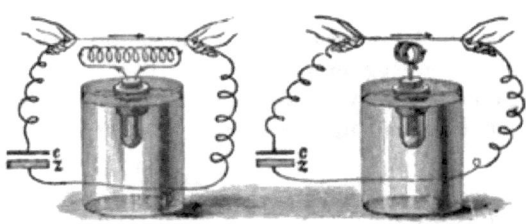

Fig. 164.

Reverse the direction of the current above and the deflection is reversed.

Second Law of Currents. — *Angular currents tend to become parallel and flow in the same direction.*

Experiment 154. — Remove the primary coil from the secondary coil (Fig. 169), send a current through the former, and hold one of its ends near to one end of the coil of the floating battery, as in Figure 165, in such a manner that the current will flow in the same direction in the ends presented to each other. The coils attract one another like two magnets in accordance with the First Law of Currents. Present the same end of the coil to the other end of the floating battery coil. Now the currents in the two ends flow in opposite directions, and the coils repel each other.

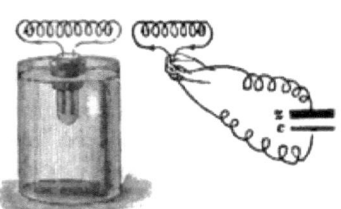

Fig. 165.

Experiment 155. — Observe that at one end of the floating battery coil the current revolves in the direction that the hands of a watch move, and at the opposite end it revolves in a direction contrary to the movement of the hands of a watch. Bring the north pole of a bar-magnet near that end of the coil where the motion of the current corresponds to the movement of the hands of a watch. They attract one another; but if the same end of the coil is approached by the south pole of the magnet, repulsion follows.

Hence, that is the south pole of a helix where the current corresponds to the motion of the hands of a watch, S, and that is the north pole where the current is in the reverse direction, N. But the important lesson derived from these latter experiments is, that *coils through which currents are flowing behave toward one another, or toward a magnet, in many respects as if they were magnets.*

171. Ampère's Theory of the Magnet.—Facts like those which we have just studied led Ampère to devise a theory for the explanation of magnetism. Little credence is given to this theory by electricians; nevertheless a slight acquaintance with it is of great service to the beginner in aiding him to picture to himself how certain phenomena occur. Ampère was led to suppose that something like an infinite number of currents invests at all times every piece of steel, iron, and other magnetizable substance. That in a magnetizable bar of steel or iron these currents are all parallel with one another, and we have the combined effects (*i.e.* of attraction or repulsion) of all the currents. When a magnet, having all its currents parallel, is brought near to an unmagnetized piece of iron or steel in which the currents have no common direction, the former induces magnetism in the latter, *i.e.* it causes the currents of the latter to become parallel with its own, in accordance with the Second Law of Currents. For convenience we may call the hypothetical currents *Ampèrian currents*

Fig. 166.

This ingenious theory will enable us to understand how the core of the electro-magnet is magnetized. The *real* currents circulating in the wire outside cause the Ampèrian currents to become parallel with them, and as both flow in the same direction as represented in Figure 166, we have, in the electro-magnet, the combined effect of both sets of currents.

172. Lines of Magnetic Force; Magnetic Field.

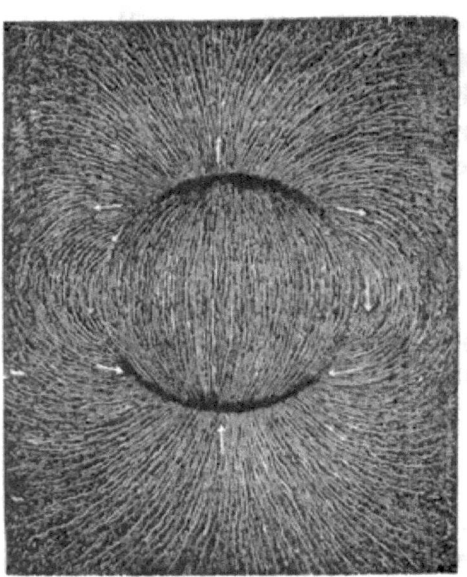

Fig. 166a.

Experiment 156. — Support a small pane of window glass on a table, by placing under the glass near its angles four slices of cork about one-eighth of an inch thick. Beneath the center of the glass on the table place a circular disk of magnetized steel. Sift iron turnings upon the upper face of the glass through a fine wire sieve. Gently tap the glass at convenient points with the end of a lead-pencil. The filings arrange themselves in lines radiating from either pole, and form graceful curves from pole to pole, as represented in Figure 166a. These represent what are called *lines of magnetic force*. They represent the results of the combined action of the two poles.

A magnet seems to be surrounded by an atmosphere of magnetic influence called the *magnetic field*. A body brought within the limit of its influence is said to be *within the field of the magnet*.

173. The Earth is a Magnet. — A *dipping-needle* is so supported that it can revolve in a vertical plane. Indifferent equilibrium is first established in the steel needle, so that if placed in a horizontal (or any other) position it will rest in that position. Then it is strongly magnetized.

Afterward it will take the horizontal position only at the magnetic equator of the earth.

Experiment 157.— Place a dipping-needle over the + pole of a bar-magnet (Fig. 167). The needle takes a vertical position with its — pole down. Slide the supporting stand along the bar; the — pole gradually rises until it reaches the middle of the bar, where it becomes horizontal. Continue moving the stand toward the — pole of the bar;

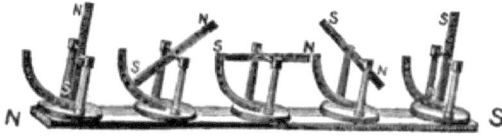

Fig. 167.

after passing the middle of the bar the + pole begins to dip, and the dip increases until the needle reaches the end of the bar, when the needle is again vertical with its + pole down.

If the same needle is carried northward or southward along the earth's surface, it will dip in the same way as it approaches the polar regions, and be horizontal only at or near the equator.

Experiment 158.— Suspend a small magnetized cambric needle by a fine thread at its center and carry it around the disk (Fig. 166). The needle passes through all the phases stated above, so that we may fancy the disk to be the earth, and study therefrom, in a general way, the changes that the needle undergoes, as it is carried around the earth in a northerly or southerly direction.

174. Magnetic Poles of the Earth. — Those points on the earth's surface where the dipping-needle stands vertical are the magnetic poles of the earth. A point was found a little northwest of Hudson's Bay, in latitude 70° 5' N., and longitude 96° 45' W., by Sir James Ross, in the year 1832, where the dipping-needle lacked only one-sixtieth of a degree of being vertical. The same voyager subsequently reached a point in Victoria Land where the needle with its poles reversed lacked only 1° 20' of being vertical.

The magnetic poles are not, however, fixed objects that can be located like an island or cape, but are constantly changing. They appear to swing, somewhat like a pendulum, in an easterly and westerly direction, each swing requiring centuries to complete it. The north magnetic pole is now on its westerly swing.

175. Variation of the Needle. — Inasmuch as the magnetic poles of the earth do not coincide with the geographical poles, it follows that in most places the needle does not point due north and south. The angle which the needle makes with the geographical meridian is known as the *angle of declination*. This angle differs at different places.

176. Inclination or Dip of the Needle. — The angle that a dipping-needle makes with a horizontal line is called its *inclination* or *dip*. A line drawn around the earth connecting those places where there is no dip would represent the *magnetic equator*.

Experiment 159. — Place the dipping-needle on a horizontal surface, apart from any iron (such as nails, etc.), and so that the plane of rotation of the needle will be in the magnetic meridian, and ascertain from the divided arc (approximately, at least) the dip at the place where you live.

EXERCISES.

1. Stretch a string between two pins stuck in a table, so that it will lie in the geographical meridian, *i.e.* in the direction of the North Star. On this string set the stand holding a magnetic needle about 6 inches long. Determine whether there is any magnetic declination at the place where you are, and, if so, in what direction it is.

2. What is the declination and dip at your place of residence?

Let A (Fig. 168) represent a magnetic pole and B the North Star. It will be seen that there is a position in which the needle will point due north. A line passing around the earth through the two magnetic poles, connecting those places where the needle points due north, is called a *line of no variation*.

3. Take a map of the United States and draw on it a pencil line, starting at a point on the Atlantic coast where the two Carolinas meet; continue it a little west of Pittsburg, Pa., and through lakes Erie and Huron, and this line will represent very nearly the line of no variation at the present time. It is slowly moving westward. At places in the United States east of this line the + pole of the needle points west of north, *e.g.* the New England States and New York; but most of the States lie west of this line, so in them the needle points east of north. At Harvard University, in Cambridge, Mass., in 1887, the declination was 11.87° W. of N.; in 1872 it was 0.7°. In 1880 the declination at Halifax, N.S., was 20.3° W. of N.; at San Francisco it was 16.52° E. of N.

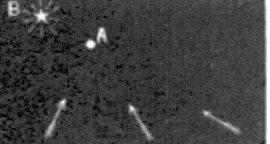

Fig. 168.

Section XIII.

CURRENT AND MAGNETIC ELECTRIC INDUCTION.

177. Description of Apparatus. — A (Fig. 169) is a short coil of coarse wire (*i.e.* the wire which it contains is comparatively short), and has, of course, little resistance. B is a long coil of fine wire having high resistance. Coil A is in circuit with two Bunsen cells in arc. This circuit we call the *primary circuit*, the current in this circuit the *primary* or *inducing current*, and the coil the *primary coil*. Another circuit, having in it no battery or other means of generating a current, contains coil B and a galvanoscope with an astatic needle. This circuit is called the *secondary circuit*, the

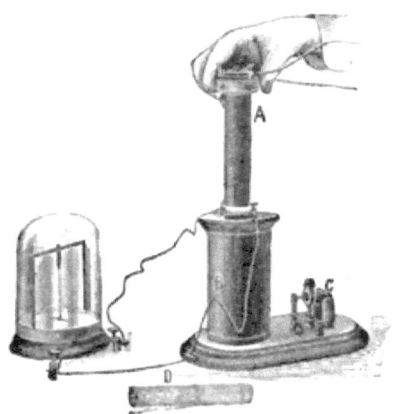

Fig. 169.

coil the *secondary coil*, and the currents which circulate through this circuit are called *secondary* or *induced currents*.

Experiment 160. — After all the connections are made, and a current is established in the primary circuit, and the galvanoscope needle is brought to zero, lower the primary coil quickly into the secondary coil, watching at the same time the needle of the galvanoscope to see whether it moves, and, if so, in what direction. Simultaneously with this movement is a movement of the needle, showing that a current must have passed through the secondary circuit. Let the primary coil rest within the secondary, until the needle comes to rest. After a few vibrations the needle settles at zero, showing that the secondary current was a temporary one. Now, watching the needle, quickly pull the primary coil out; another deflection in an opposite direction occurs, showing that a current in an opposite direction is caused by withdrawing the coil. Just how the necessary condition (*i.e.* E.M.F.) for an electric current is brought about we do not know; but we do know that it is done under the *influence* of the primary current (hence the process is called *induction*) and *at the expense of mechanical energy*.

Fig. 170.

Experiment 161. — Place the primary coil within the secondary. Open the primary wire at some point and then close the circuit (Fig. 170) by bringing in contact the extremities of the wires. A deflection is produced. As soon as the needle becomes quiet, break the circuit by separating the wires; a deflection in the opposite direction occurs.

On introducing the primary coil into the secondary, and on closing the primary circuit, currents are induced in the reverse direction in the secondary circuit that the primary current has; on the withdrawal of the primary coil, or on breaking the primary circuit, the induced current generated is in the same direction as of that of the primary current.

Experiment 162.— Introduce the bundle, D (Fig. 169), of soft iron wires, called the *core*, into the primary coil, and make and break the primary circuit as before. The deflections are now very much increased.

Experiment 163.— Substitute a person for the galvanometer in the secondary circuit, the person grasping some metallic handles made for the purpose and used as electrodes. The person experiences at the instant of making and breaking a peculiar sensation in his wrists and arms, called a *shock*.

Experiment 164.— Introduce into the primary circuit the automatic make-and-break piece C (Fig. 169). Remove the core from the primary coil. Let a person grasp the electrodes of the secondary circuit. This person experiences a series of shocks which seem to him almost, if not quite, continuous. These shocks can be intensified to suit the pleasure of the person who is receiving them, by gradually lowering the core into the primary coil. But no temptation to fun should lead the experimenter to be so cruel as to drop the core into the coil suddenly.

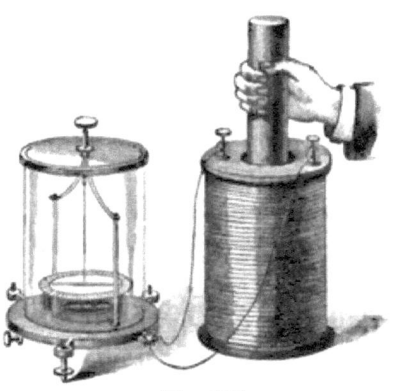

Fig. 171.

Experiment 165.— Reflecting that you have found hitherto a coil of wire having a current passing through it acting as a magnet, you have now an opportunity to try the converse, *i.e.* to see whether a magnet may not take the place of a current-bearing coil. Introduce suddenly a bar-magnet (Fig. 171) into the secondary coil, as in Experiment 160. A deflection is produced; withdraw it and an opposite deflection occurs.

Laws of Induced Currents: The general laws of induced currents are summed up in the following table : —

INDUCTOR.	INVERSE INDUCED CURRENT.	DIRECT INDUCED CURRENT.
A *magnet* . . .	Approaching.	Receding.
A *current* . . .	Approaching. Beginning. Increasing in strength.	Receding. Stopping. Diminishing in strength.

178. Extra Currents. — Pupils, while handling the naked electrodes of a battery having an electro-magnet or other coil in the circuit, at the instant of dropping or taking hold of the electrodes frequently experience slight shocks. This is due to what are called *extra* currents induced in the battery circuit itself at the instants of making and breaking. As the battery current advances or retires through the wire, each convolution of wire acts inductively upon the neighboring convolutions, in a manner similar to that of the primary coil upon the secondary. The sparks invariably attending the touching and separating of electrodes, *e.g.* those seen at the make-and-break piece C (Fig. 169), are produced by extra currents.

179. Ruhmkorff's Induction Coil. — Figure 172 represents, in diagram, an ideal induction coil. A A is the core around which is wound the primary wire. Outside of the whole is the secondary coil. The directions of the several currents are indicated by arrows at the instant the primary circuit is closed at *b* in the automatic piece *cd*. The condenser B B was the important addition made by Ruhmkorff.

It consists of two sets of layers of tin-foil separated by paraffine paper; the layers are connected alternately with one and the other pole of the battery, as the figure shows, so that they serve as a sort of expansion of the primary wire. When the circuit is broken, the extra current would jump across at *b*, and would vaporize the points of contact, and form a bridge with the vapor of metal that would prolong the time of breaking. But, when the condenser is attached, the extra current finds an escape into

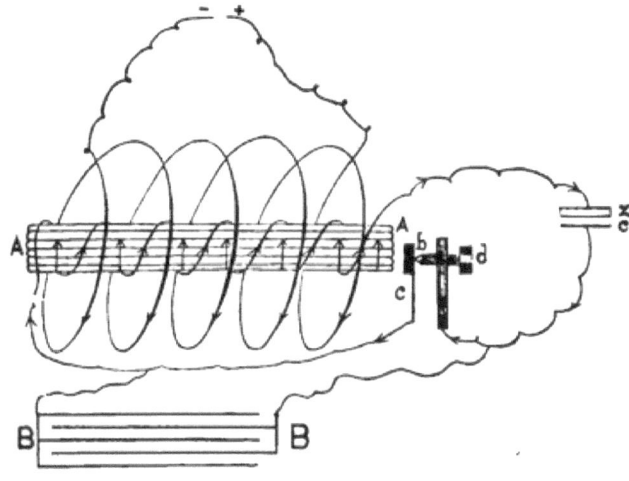

Fig. 172.

it easier than to jump across at *b*, so the vaporizing of the contact is avoided, and the time of breaking being much shortened, the secondary current is much more intense.

Experiment 166. — Connect a battery of two Bunsen cells, in arc, with a Ruhmkorff coil (Fig. 173). Bring the electrodes of the secondary coil within one-fourth of an inch to one inch of each other, according to the capacity of the instrument. A series of sparks in rapid succession pass from pole to pole.

Experiment 167. — Introduce a *Geissler tube*, A, into the secondary circuit. These tubes contain highly rarefied gases of different kinds. Platinum wires are sealed into the glass at each end to conduct the electric cur-

Fig. 173.

rent through the glass. The sparks become diffused in these tubes so as to illuminate the entire tubes with an almost continuous glow. Observe that the electrodes are separated from each other much more widely than would be admissible in air of ordinary density, showing that rarefied gases offer less resistance than dense gases. Gases have been so highly rarefied, however, that an electric current would not pass. This shows that a *material* conductor and one *of sufficient density* is absolutely necessary for the passage of a current.

180. Electric Motor.

Experiment 168. — This experiment will require two separate batteries. Join one battery to a small Ruhmkorff coil, and connect its secondary coil with the apparatus represented in Figure 174, introducing the wires at the binding screws, c and d. Join the wires of the other battery with the same instrument, inserting the wires at the binding screws, a and b. The first battery in conjunction with the coil causes induced currents to enter this instrument and pass through the Geissler tube, A. The other battery causes the tube to rotate. In a darkened room the appearance is that of a luminous wheel of great beauty having many spokes. Various optical illusions attend the experiment, which make it very attractive.

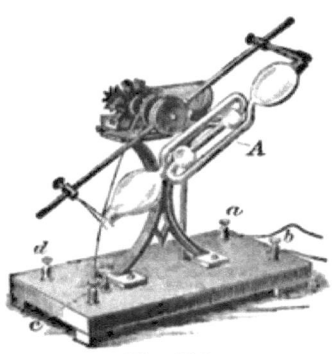

Fig. 174.

The instrument used is one form of an *electric motor*. An electric motor is a device for transforming the energy of an electric current into mechanical energy, *i.e.* into motive power. It is usually accomplished through the use of electro-magnets, and hence a motor is frequently called an *electro-magnetic engine*. Electric motors of great power have been constructed, and are successfully used for propelling railway cars, etc.

181. Characteristics of Induced Currents. — The student cannot have failed to observe that induced electricity has a power for penetrating a non-conductor far superior to that of primary currents.

The former can penetrate the air passing through it from electrode to electrode, at distances varying from one-hundredth of an inch to three feet in the largest induction coils. They can perforate cardboard, panes of glass, and produce various other mechanical effects. They may be so intense as to produce instantaneous death. On the other hand, it would require the E.M.F. of several thousand voltaic cells connected, in series, to furnish sufficient power to penetrate the air so as to maintain a current when the electrodes are separated only one-hundredth of an inch.

Section XIV.

DYNAMO-ELECTRIC MACHINES.

182. A Simple Dynamo and the Gramme Dynamo.

Experiment 169. — Take the secondary coil of the induction coil apparatus (Fig. 169), place within it the core of iron wires. Introduce into circuit with this coil a galvanoscope with an astatic needle. Take a powerful compound horseshoe; suspend it in a vertical position with the poles downward. Move the coil back and forth under and near to the magnet, so that the core will come alternately under each pole. Deflections alternating in direction show the production of induced currents.

The student should look thoughtfully at this contrivance, because he has before him a *dynamo-electric machine* in its simplicity. It consists, like all the more complicated machines, of these two *essential parts*, viz. (usually) a long coil containing an iron core, constituting an *armature*, and a powerful magnet (either a permanent steel magnet, or, more frequently, because more powerful, an electro-magnet) called the *field magnet*. The method by which currents are generated in this contrivance and in all dynamos is the same, viz. *by the movement of an armature within the field of an electro-magnet.*

Much more than the above it is not important that the general student should know. Matters of detail differ widely in different machines, and the student is not supposed to be especially interested in any particular machine. To give a full and intelligible description of any machine in a single page is not an easy matter. For the benefit of the more ambitious students, we submit the following condensed description of the Gramme dynamo. Its armature, *ns* (Fig. 175), consists of a ring composed of a bundle of soft iron wires (better shown in Figure

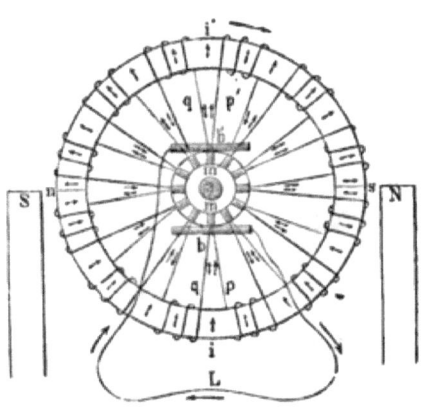

Fig. 175.   Fig. 176.

177, Plate III.) surrounded by what is virtually an endless coil of wire. The wire, however, is wound in sections separated by suitable partitions, and the wire of each section carried to and connected electrically with a copper plate on the axle *mm*. The several copper plates (as many as there are sections) are insulated from one another. (To enable the pupil better to understand the method of winding, making connections, etc., the author has prepared a model (Fig. 176) of this machine, which will furnish at a glance information respecting the method of winding, making connections, etc., which no book can do.) A horseshoe magnet NS (only a portion of which is shown in the cut) is so placed that one-half of the ring is under the influence of the N-pole, and the other half under that of the S-pole. Suppose the ring to rotate in the direction of the arrow; then every point of the iron core, as it comes opposite a given point of the magnet, will successively become a pole of opposite name, while the points i and i' are the neutral points.

Plate III.

Fig. 179.

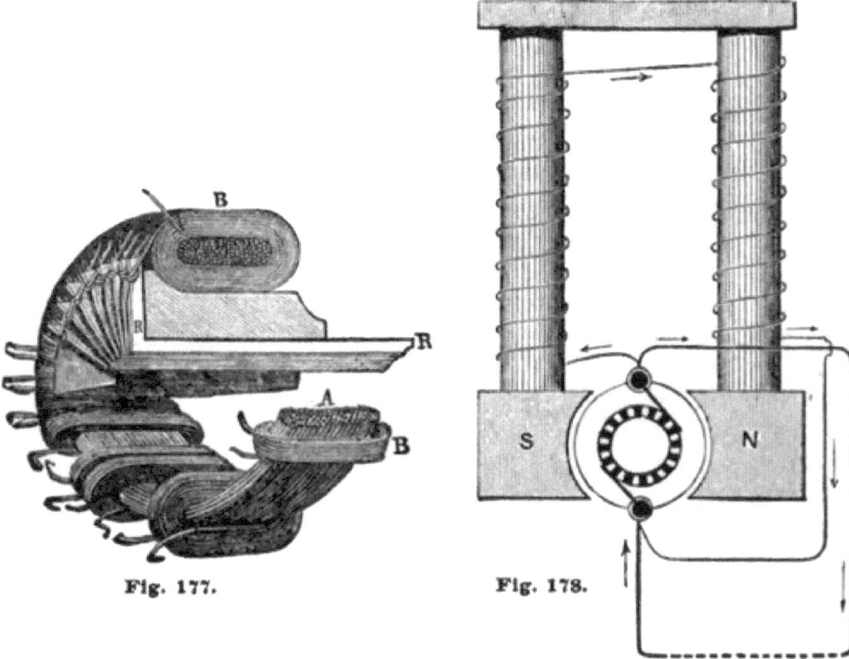

Fig. 177. Fig. 178.

If we imagine the core to be divided at the points n and s, we have two semicircular magnets whose north poles and whose south poles respectively face one another. In the two mutually facing poles on either side, the Ampèrian currents must be in opposite directions. Now an attentive study of this ideal diagram, in the light of what you have previously learned respecting the generation of induced currents, will enable you to see that as the ring armature rotates, the corresponding advance of the induced poles of the ring will induce currents in the wire in such a manner that all the coils which at any given moment are in the semicircle next one of the magnet poles (say the North) are traversed by a current in one direction. Similarly, the semicircle formed by the coils immediately approaching, or immediately receding from the South pole are at the same time traversed by a current in the opposite direction. The result is that currents in the lower half tend *toward* the point m on the axis, and in the upper half *from* point m'. So long as the leading-out wires from these points are open, these currents have no outlet, and consequently oppose and neutralize one another. But if the points m and m' are connected by a wire L, we shall have a *constant and non-alternating current* flowing through the wire from m to m'. The contact at these points is made by means of brushes of thick wire. These press on the contact pieces, and make practically a constant connection with the two halves of the circuit.

Inasmuch as an electro-magnet may be made a much more powerful magnet than a permanent magnet, it is now extensively used as the inducing or the so-called *field magnet*. Such a machine is called a *dynamo-electrical machine*, or often more briefly a *dynamo*. Figure 178, Plate III., represents such a machine. E E is the stationary field magnet, A, the moving armature, and N and S large pole-pieces brought as near as practicable to the armature and partially encircling it. When the machine is at rest, there are no currents; but when the armature is in motion, the residual magnetism (a small portion of which is always retained by soft iron after it has been magnetized) induces at first a weak current in the wire of the armature; but as a portion of this current is carried by means of a shunt wire l through the coil of the field magnet, and magnetizes the core more strongly, the current in both the shunt l and the main wire L quickly reaches its maximum.

By permission of the United States Electric Lighting Company we introduce a cut (Fig. 179, Plate III.), of the American dynamo called the Weston. It will be seen that in this machine a powerful field magnet is placed on each side of the revolving armature. A steam-engine communicates motion to the dynamo by means of a belt passing over the circumference of the wheel W, and causes the armature, which is on the axle of this wheel, to revolve.

183. The Dynamo as an Electric-motor. — If, instead of expending mechanical energy, such as that of a steam-engine, etc., in rotating the armature of a dynamo, a current from another dynamo (or other source) is sent through the coil of its armature, the armature will rotate under the action of the electric energy, and the dynamo thus becomes an *electric-motor*. In the generating dynamo mechanical energy is transformed into electric energy; in the receiving dynamo (used as a motor) the electric energy is transformed again into mechanical energy. A series of dynamos (only limited in number by the loss of energy by waste) might be so connected that transformation in each is the reverse of that in the preceding.

184. Uses of Dynamos. — We live at the interesting epoch when dynamos are being rapidly introduced for purposes of electric lighting, electroplating, motive power, telegraphy, charging storage batteries, etc., supplanting to a large extent other instrumentalities and branches of industry, much as sixty years ago the locomotive commenced its displacement of the stage coach.

185. Transmission of Electric Energy. — One of the most important projects which is enlisting the attention of electricians at the present time is to devise some efficient means of economically transforming, by means of dynamos, some of the wasting energies of nature, such, for example, as that of waterfalls, into electric energy, and in this convenient form transferring the energy through wires to distant and available places, such as large cities, where it may be transformed by lamps into heat and light, or by electric-motors into mechanical energy for doing almost any kind of work. The project is theoretically possible. One of the principal practical difficulties is that of safely, and without great waste, transmitting currents of great magnitude long distances through conductors such as are now in use. In many ways electric energy is one of the most convenient forms of energy; hence its desirability for propelling street cars, for operating light machinery, etc. It is apparent that if this form of energy could somehow be, as it were, bottled up or stored in large quantities in a small space, so that it could be transported easily to places where it is needed, it would be a valuable achievement. This is in a measure practicable through the agency of the so-called "storage batteries."

186. Storage Batteries. — The storage battery is virtually an electrolysis apparatus, having instead of two platinum electrodes two lead plates coated with red lead (Pb_3O_4) with a layer of paper or cloth between, the whole suspended in dilute sulphuric acid. (See directions for making storage batteries in the author's Physical Technics, page 122.) When these electrodes are connected with a powerful voltaic battery, or, better, with a dynamo, the $+$ electrode becomes peroxydized (PbO_2) by the oxygen liberated by electrolysis, while the $-$ electrode is deoxydized by the hydrogen liberated. In other words, the energy of the current is transformed into the potential energy of chemical affinity. Note that *it is an electrical storage of energy, not a storage of electricity,* — two very different things. When these chemical changes have progressed as far as possible the battery is said to be charged. These plates may remain for many days in this condition, if the circuit is left open, and may be transported long distances and used in the same way and for the same purposes that any powerful voltaic battery can be used. Storage cells may be combined the same as voltaic cells (which in fact they are after charging), and with similar results. Some idea of the capacity of these cells may be formed from the following estimate. In a cell whose interior dimensions are eight inches square and four inches deep, there can be stored up energy sufficient to furnish one-half of a horse-power working for an hour.

Section XV.

USEFUL APPLICATIONS OF ELECTRIC ENERGY. — ELECTRIC LIGHT.

The applications of electric energy to industrial uses are so numerous and varied that the limits of an ordinary text-book on general Physics can do little justice to the subject, and, indeed, a description of the various appliances in use is of a too technical character to come properly within the scope of a general high-school course. Public libraries are now well provided with popular works relating to every industrial application. Students may consult with profit such books as Prescott's The

Telegraph and Telephone, Dolbear's The Telephone, Urquhart's Electro-plating, S. P. Thompson's Dynamo-Electric Machinery, and Sawyer's Electric Lighting.

187. Electric Light: Voltaic Arc. — If the terminals of wires from a powerful dynamo or galvanic battery are brought together, and then separated 1 or 2^{mm}, the current does not cease to flow, but volatilizes a portion of the terminals. The vapor formed becomes a conductor of high resistance, and remaining at a very high temperature produces intense light. The light rivals that of the sun both in intensity and whiteness. The heat is so great that it fuses the most refractory substances, including even the diamond. Metal terminals quickly melt and drop off like tallow, and thereby become so far separated that the electro-motive force is no longer sufficient for the increased resistance, and the light is extinguished. Hence, pencils of carbon (prepared from the coke deposited in the distillation of coal inside of gas retorts), being less fusible, are used for terminals.

Fig. 180.

For simple experiments, these pencils may be held in forceps (Fig. 180) at the ends of two brass rods, to which the battery wires are attached. These rods slide in brass heads, A and B, supported by insulating pillars, so that the distance between the carbon points may be regulated.

The light is too intense to admit of examination with the naked eye; but if an image of the terminals is thrown on a screen by means of a lens, or a pin-hole in a card, an arch-shaped light is seen extending from pole to pole, as shown in Figure 181. This light has received the name of the *voltaic arc*. The larger portion of the light, however, emanates from the tips of the two carbon terminals, which are heated to an intense whiteness, but some emanates from the arc. The + pole is hotter than the − pole, as is shown by its glowing longer after the current is stopped. The carbon of the + pole becomes volatilized, and the light-giving particles are transported from the + pole to the − pole, forming a bridge of luminous vapor between the poles. What we see is not electricity, but *luminous matter*.

Fig. 181.

The light of the ordinary street arc-lamp has an intensity varying from one to two thousand candle-power, or the combined intensity of from fifty to a hundred ordinary gas-lights. To sustain such a light, about one horse-power per lamp must be applied at the dynamo.

188. Electric Lamp. — It is apparent that the + pole is subject to a wasting away; so also the − pole wastes away, but not so fast. At the point of the former a conical-shaped cavity is formed, while around the point of the latter warty protuberances appear. When, in consequence of the wearing away of the + pole, the distance

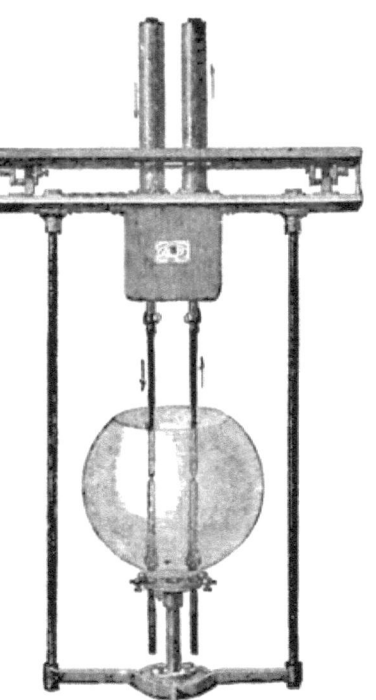

Fig. 182.

between the two pencils becomes too great for the electric current to span, the light goes out. Numerous self-acting regulators for maintaining a uniform distance between the poles have been devised. Such an arrangement (Fig. 182) is called an *electric lamp*. The movements of the carbons are accomplished automatically by the action of the current itself.

The difference between the arc-lamps of the various inventors is a difference in the mode of adjusting or "feeding" the carbons. We give below the plan of the

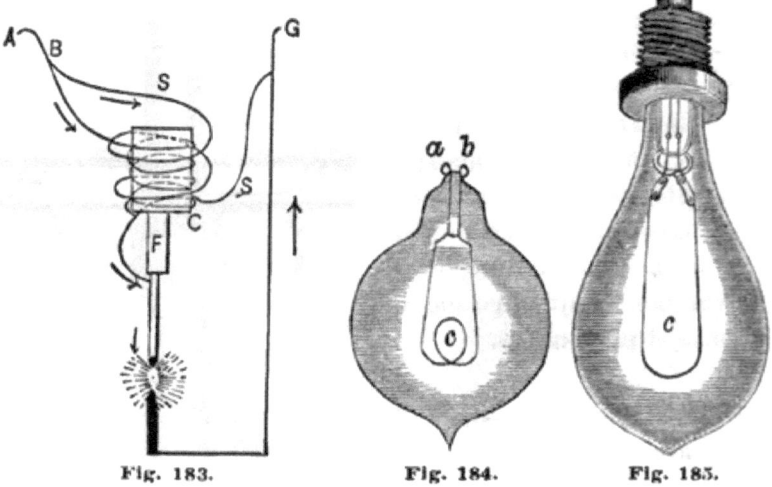

Fig. 183.   Fig. 184.   Fig. 185.

189. Brush Lamp. — The current, entering at A (Fig. 183), divides at B into two branches which pass around the bobbin C in opposite directions, one branch being a coarse wire of low resistance and in the same circuit as the carbons, and the other branch SS being a shunt of high resistance, connecting the terminals B and G. Inside the bobbin is a soft iron core, F, which is attached to the upper carbon. When a current passes through the two branch circuits on the bobbin C, they tend to magnetize the core in opposite directions, but the resistances and number of turns in the two circuits are so proportioned that the magnetic field due to the low resistance branch is the stronger, and the core F is therefore

ELECTRIC LIGHT.

drawn up into the bobbin, lifting the upper carbon and establishing the arc. Should the carbons become too widely separated the resistance of the arc, and consequently of the coarse wire circuit on C, increases, diminishing the current in C and increasing that in the shunt S. The field due to the shunt is therefore strengthened, and that due to the coarse wire diminished, allowing the core F to fall slightly, bringing the carbons nearer together. By the device of the two opposing fields, due to the coils on C being wound in opposite directions, the feeding of the lamp is done automatically, and the actual distance of the two carbons varies but little.

190. Incandescent Electric Lamps. — The incandescent (or "glow") light is produced by the heating of some refractory body to a state of incandescence by the passage of an electric current, as, for example, the light given off by heated platinum in Experiment 120. Platinum is little used for this purpose on account of its liability to melt. Carbon filaments are now exclusively used in incandescent lamps. In the Swan lamp (Fig. 184) a filament of carbonized cotton, twisted into a sort of curl, is attached at its ends to two little platinum wires, a and b, which have previously been sealed into the neck of the glass bulb. The filament of the Edison lamp (Fig. 185) is carbonized bamboo. It is essential that the oxygen of the air be removed from these bulbs, otherwise the carbons would be quickly burned out; hence very high vacua are produced in the bulbs with a mercury pump.

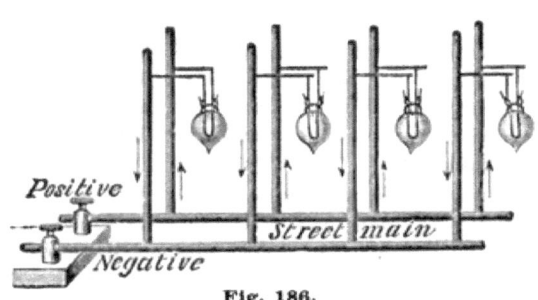

Fig. 186.

An Edison 16 candle-power lamp has a resistance (when hot) of about 140 ohms, the difference of potential at its terminals is about 100 volts, and it requires a current of 0.75 ampère. Each lamp consumes about one-tenth of a horse-power.

Incandescent lamps are usually introduced into the circuit in multiple arc (Fig. 186), the current being equally divided by properly regulating the resistance between all the lamps in the circuit.

Section XVI.

USEFUL APPLICATIONS OF ELECTRICITY CONTINUED. — ELECTROTYPING AND ELECTROPLATING.

191. Electrotyping.—This book is printed from electrotype plates. A molding-case of brass, in the shape of a shallow pan, is filled to the depth of about one-quarter of an inch with melted wax. A few pages are set up in common type, and an impression or mold is made by pressing these into the wax. The type is then distributed, and again used to set up other pages. Powdered plumbago is applied by brushes to the surface of the wax mold to render it a conductor. The case is then suspended in a bath of copper sulphate dissolved in dilute sulphuric acid. The — pole of a galvanic battery or dynamo machine is applied to it; and from the + pole is suspended in the bath a copper plate opposite and near to the wax face. The salt of copper is decomposed by the electric current, and the copper is deposited on the surface of the mold. The sulphuric acid appears at the + pole, and, combining with the copper of this pole, forms new molecules of copper sulphate. When the copper film has acquired about the thickness of an ordinary visiting card, it is removed from the mold. This shell shows distinctly every line of the types or engraving. It is then backed, or filled in, with melted type-metal, to give firmness to the plate. The plate is next fastened on a block of wood, and thus built up type-high, and is now ready for the printer. (For full directions which will enable a pupil to electrotype in a small way, see the author's Physical Technics.)

192. Electroplating. — The distinction between electroplating and electrotyping is, that with the former the metallic coat remains permanently on the object on which it is deposited, while with the latter it is intended to be removed. The processes are, in the main, the same. The articles to be plated are first thoroughly cleaned and suspended on the

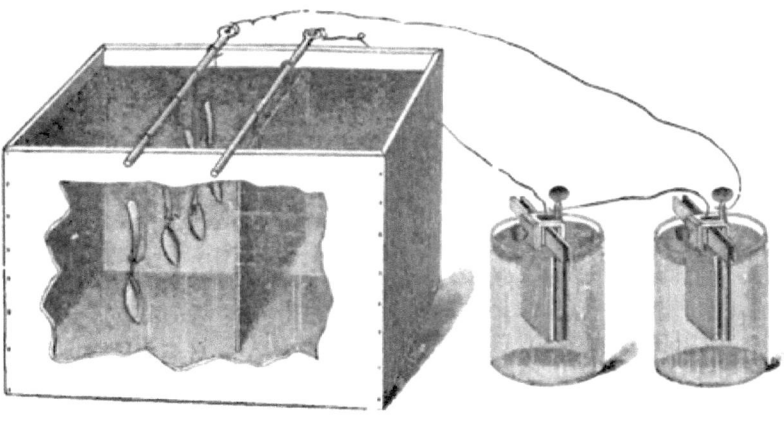

Fig. 187.

— pole of a battery, and then a plate of the same kind of metal that is to be deposited on the given articles is suspended from the + pole (Fig. 187). The bath used is a solution of a salt of the metal to be deposited. The cyanides of gold and silver are generally used for gilding and silvering. Many of the base metals require to be electro-coppered first, in order to secure the adhesion of the gold or silver. The magneto-electric machine has almost completely replaced the voltaic battery for electrotyping and electroplating purposes.

Section XVII.

USEFUL APPLICATIONS OF ELECTRIC ENERGY CONTINUED. — TELEGRAPHY.

193. The Telegraph. — The word *telegraph*, literally, signifies *to write far away*. In its broadest sense it embraces all methods of communicating thought with great speed to a distance, by means of intelligible characters, sounds, or signs; but usually it is applied only to electrical methods.

First, it should be understood that, instead of two lines of wire, one to convey the electric current far away from the battery, and another to return it to the battery, if the distant pole is connected with a large metallic plate buried in moist earth, or, still better, with a gas or water pipe that leads to the earth, and the other pole near the battery is connected in like manner with the earth, so that the earth forms about one-half of the circuit, there will be needed *only one wire* to connect telegraphically two places that are distant from each other. Furthermore, *the resistance offered by the earth to the electric current is practically nothing*; so that, disregarding the resistance of the ground connections, there is a saving of one-half the wire and one-half the resistance, and consequently of one-half the battery power.

Let B, Figure 188, Plate IV., represent the message sender, or operator's key; Y, the message receiver. It may be seen that the circuit is broken at B. Let the operator press his finger on the knob of the key. He closes the circuit, and the electric current instantly fills the wire from Boston to New York. It magnetizes a; a draws down the lever b, and presses the point of a style on a strip of paper, c, that is drawn over a roller. The operator ceases to press upon the key, the circuit is broken, and instantly b is raised from the paper by a spiral spring, d. Let the operator press upon the key only for an instant, or long enough to count one: a simple *dot* or indentation will be made in the paper. But if he presses upon the key long enough to count three, the point of the style will remain in contact with the paper the same length of time; and, as the paper is drawn along beneath the point, a short straight line is produced. This short line is called a *dash*. These dots and dashes constitute the *alphabet of telegraphy*. For instance, a part of a message, "man is in," is represented as printed in telegraphic characters on the strip of paper. The Roman letters above interpret their meaning.

Plate IV.

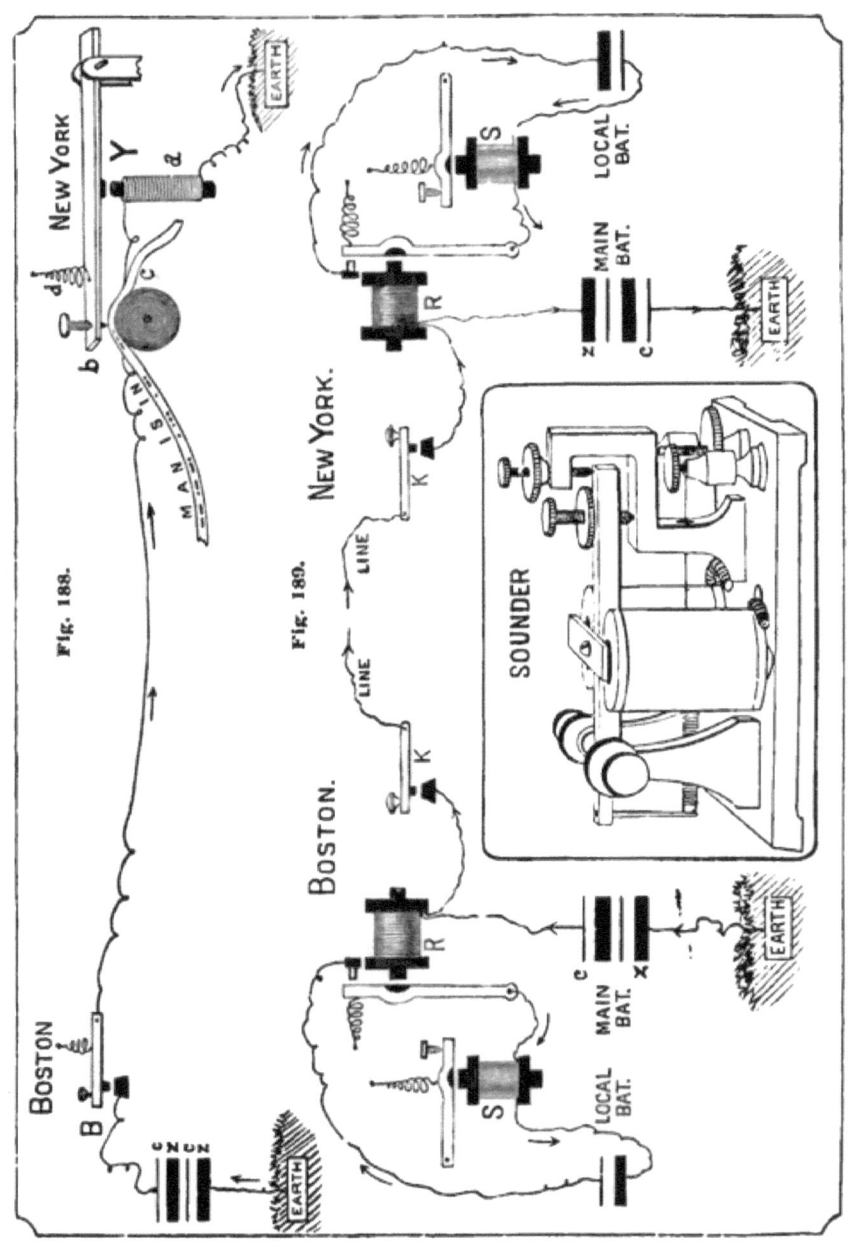

194. The Sounder. — If the strip of paper is removed, and the style is allowed to strike the metallic roller, a sharp click is heard. Again, when the lever is drawn up by the spiral spring, it strikes a screw point above (not represented in the figure), and another click, differing slightly in sound from the first, is heard. A listener is able to distinguish dots from dashes by the length of the intervals of time that elapse between these two sounds. Operators generally read by ear, giving heed to the clicking sounds produced by the strokes of a little hammer. A receiver so used is called a *sounder*, a common form of which is represented in the lower central part of Plate IV.

195. The Relay and the Repeater. — The strength of the current is diminished, of course, as the line is extended and the number of instruments in the circuit is increased. Hence, a current that would move the parts of a single sounder audibly, on a short line, would not move the same parts of many sounders on a long line with sufficient force to render the message audible. Resort is had to *relays* and *repeaters*.

In Figure 180, Plate IV., the letter R represents a relay and S a sounder. Suppose a weak current arrives at New York from Boston, and has sufficient strength to attract the armature of the relay at that station. This, as may be seen by examination of the diagram, will close another short circuit, called the *local circuit*, and send a current from a *local battery* located in the same office through the sounder at that station. The sounder, being operated by a battery in a circuit of only a few feet in length, delivers the message audibly. If it is desired that the message should go beyond New York,— for instance, to Philadelphia, — then we have only to suppose the local line at New York to be lengthened so as to extend to Philadelphia, and a powerful line battery to be substituted for the small local; then the message that leaves Boston will be shifted from one circuit to the other at New York, and be delivered in Philadelphia without the intervention of any operator on the route. In this case a relay is called a *repeater*. The electro-magnets in relays are wound with long, thin wire, while those of sounders are wound with short, large wire. The main battery consists of many cells in series. It may be located at either terminus, but it is generally split in halves, and one half placed at each terminus.

In the diagram, the circuit is represented as open at both keys. When the line is not in use, the circuit ought always to be left closed, by means of switches connected with the keys (not represented in the diagram), so that when the line is not "at work" an electric current is constantly traversing the wire. Sending a message, consequently, consists in interrupting this current by means of a key. Suppose that Boston wishes to communicate with New York. He first removes the switch on his key, which breaks the circuit and enables him to control the circuit with his key. He then manipulates his key so as to produce an understood signal, which will attract New York's attention. Every time that Boston presses on his key, every armature in his own office, and in the New York office, and at way stations, falls. Of course the message may be read at every station on the route.

TELEGRAPHIC ALPHABET.

A	B	C	D	E	F
.—	—...	.. .	—..	.	..—.
G	H	I	J	K	L
——.	....	..	—.—.	—.—	———
M	N	O	P	Q	R
——	—.	..	.....	..—.	. ..
S	T	U	V	W	X
...	—	..—	...—	.——	.—..
Y	Z	&	,	?	.
.. ..	... .	. ...	.—.—	—..—.	..——..

TELEGRAPHIC FIGURES.

1	2	3	4	5	6
.—— .	..—..	...—.	....—	———	......
	7	8	9	0	
	——..	—....	—..—	———	

Section XVIII.

USEFUL APPLICATIONS OF ELECTRIC ENERGY CONTINUED. — TELEPHONY.

196. Bell Telephone. — Figure 190 represents a sectional and a perspective view of this instrument. It consists of a steel magnet A, encircled at one extremity by a spool B of very fine insulated wire, the ends of which are connected with the binding-screws DD. Immediately in front of the magnet is a thin circular iron disk EE. The whole is enclosed in a wooden or rubber case F. The conical-shaped cavity G serves the purpose of either a mouth-piece or an ear-trumpet. There is no difference between the transmitting and receiving telephone; consequently either instrument may be employed as a transmitter, while the other serves as a receiver. Two magneto telephones in a circuit are virtually in the relation of a dynamo and a motor. The transmitter being in itself a diminutive dynamo, of course no battery is required in the circuit. Connect in circuit two such telephones, and the apparatus is ready for use.

When a person talks near the disk of the transmitter, he throws it into rapid vibration. The disk, being quite close to the magnet, is magnetized by induction; and as it vibrates, its magnetic power is constantly chang-

Fig. 190.

ing, being strengthened as it approaches the magnet, and enfeebled as it recedes. This fluctuating magnetic force will of course induce currents in alternate directions in the neighboring coil of wire. These currents traverse the whole length of the wire, and so pass through the coil of the distant instrument. When the direction of the arriving current is such as to re-enforce the power of the magnet of the receiver, the magnet attracts the iron disk in front of it more strongly than before. If the current is in the opposite direction, the disk is less attracted, and flies back. Hence, whatever movement is imparted to the disk of the transmitting telephone, the disk of the receiving telephone is forced to repeat. The vibrations of the latter disk become sound in the same manner as the vibrations of a tuning-fork or the head of a drum.

The above is a description of the original and simplest form of the Bell telephone. It is apparent that the original energy, *i.e.* that of the voice, applied at the transmitter must, during its successive transformations and especially during its transmission in the form of electric energy through large resistances, become very much enfeebled, so that when it reappears as sound, the sound is quite feeble and frequently inaudible. The first grand improvement on the original consists in introducing a battery into the circuit, and so arranging that the voice, instead of being obliged to generate currents, should be required to act only as a controlling force of a current already generated by the battery. It is evident that only a fluctuating or undulating current can produce the necessary vibrations in the disk of the receiver. The fluctuations are caused by a vary-

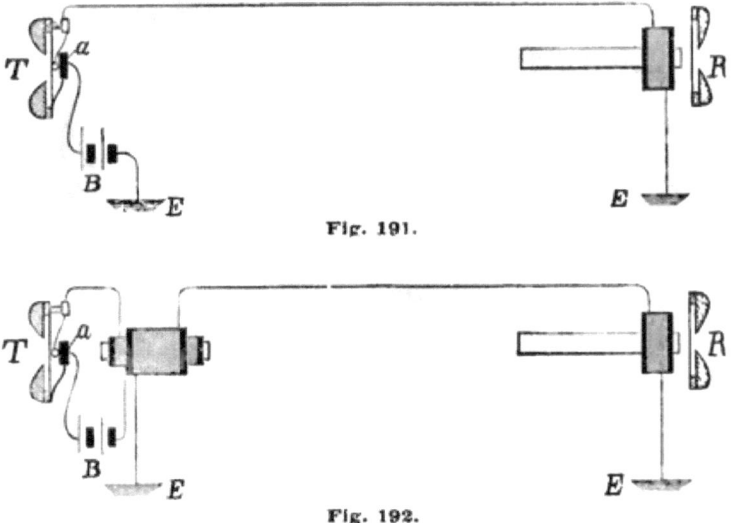

Fig. 191.

Fig. 192.

ing resistance in the circuit. The pupil must have learned by experience ere this that the effect of a loose contact between any two parts of a circuit is to increase the resistance and thereby weaken the current; but the effect of a slight variation in pressure is especially noticeable when either or both of the parts are carbon. Figure 191 illustrates a simple telephonic circuit in which are included a variable resistance transmitter T, a magneto receiver R, and a battery B. One of the electrodes, a platinum point, touches the center of the transmitter disk; the other electrode, a carbon button a, is pressed by a spring gently against the platinum point. Every vibration of the disk, however minute, causes a variation in the pressure between the two electrodes and a corresponding variation in the circuit resistance. As changes the resistance, so changes the current

TELEPHONY. 221

strength, and so consequently changes the force with which the magnet in the receiver R pulls its disk. The varying tension between magnet and disk causes the latter to vibrate and reproduce sounds.

The next improvement of considerable importance consists in the adoption of an induction coil, which, we have learned, produces a current of much greater electromotive force than is possessed by the original battery current. By its adoption we are able to converse over much longer distances, and since the battery current traverses only a local circuit, as may be seen by reference to Figure 192, a single Leclanché cell is generally sufficient to operate it. The currents induced by the fluctuating primary current traverse the line wire and generate sonorous vibrations in the disk of the receiver in the same manner as in the original telephone.

Figure 193 represents the entire telephonic apparatus required at any single station. The box A contains a small hand-dynamo, such as is represented in Figure 194. A person turning the crank F generates a current which rings a pair of electric bells G, both at his own and at a distant station, and thus attracts attention. He next takes the receiver B off the supporting hook and places it to his ear. When the weight is removed from the hook, the hook rises a little and throws the dynamo and bells out of the circuit, and at the same time introduces the receiver B, the transmitter C, and the battery D, so that the circuit stands as represented in Figure 192. The box C contains the induction coil. E is a "lightning arrester."

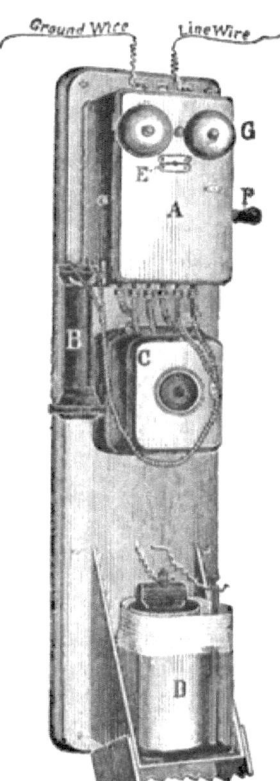

Fig. 193.

Fig. 194.

197. Microphone. — In Figure 195, A and B are buttons of carbon; the former is attached to a sounding-board of thin pine wood, the latter to a steel spring C, and both are connected in circuit with a battery and a telephone used as a receiver. The spring presses B against A, and any slight jar will cause a variation in the pressure and corresponding variations in the current strength.

Fig. 195.

By means of this instrument, called the *microphone*, any *little sounds*, as its name indicates, such as the ticking of a watch or the footfall of an insect, may be reproduced at a considerable distance, and be as audible as though the original sounds were made close to the ear.

Section XIX.

THERMO-ELECTRIC CURRENTS.

198. Heat Energy transformed directly into Electric Energy.

Experiment 170. — Insert in one screw-cup of an astatic galvanometer an iron wire, and in the other cup a copper, or better, a German silver wire. Twist the other ends of the wire together, and heat them at their junction in a flame; a deflection of the needle shows that a current of electricity is traversing the wire. Place a piece of ice at their junction; a deflection in the opposite direction shows that a current now traverses the wire in the opposite direction.

Experiment 171. — Take a strip of sheet copper about 15 inches long and three-fourths of an inch wide, and a strip of zinc of the same dimensions. Lay them one upon the other, and fold over each end upon itself for about half an inch, and hammer the joints flat, so that they shall hold together quite firmly. Then separate the two strips into a somewhat elliptical or rectangular shape, as shown in Figure 196. Cut a hole through the center of one of the strips, and pass the wire support of a magnetic needle through it.

Fig. 196.

Place the band in the magnetic meridian parallel with the needle. Direct a flame against one of the junctions, and note the deflection, and determine the direction in which the current traverses the band, *i.e.* whether the current passes from the heated junction through the copper or the zinc strip.

These currents are named, from their origin, *thermo-electric*. The apparatus required for the generation of these currents is very simple, consisting merely of bars of two different metals joined at one extremity, and some means of raising or lowering their temperature at their junction, or of raising the temperature at one extremity of *the pair* and lowering it at the other; for the electromotive force, and consequently the strength of the current, is nearly proportional to the difference in temperature of the two extremities of the pair. The strength of the current is also dependent, as in the voltaic pair, on the thermo-electromotive force of the metals employed. The following *thermo-electric series* is so arranged that if the temperatures of both junctions are near the ordinary temperatures of the air, those metals farthest removed from each other give the strongest current when combined; and the current passes, when heated at their junction, from the one first named to that succeeding it. The arrows indicate the direction of the current at the heated and the cold ends respectively. At high temperatures the current may be reversed.

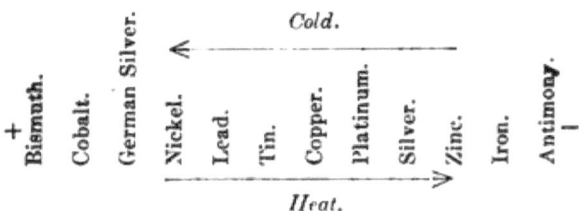

199. Thermo-electric Batteries and Thermo-pile. —

The electro-motive force of the thermo-electric pair is very small in comparison with that of the voltaic pair; hence the greater necessity of combining a large number of pairs with one another in series. This is done on the same principle, and in the same manner, that voltaic pairs are united; viz., by joining the + metal of one pair to the — metal of another. Figure 197 represents such an arrangement. The light bars are bismuth, and the dark ones antimony. If the source of heat is strong and near, one face may be heated much hotter than the other, and a current equal to that from an ordinary galvanic cell is often obtained.

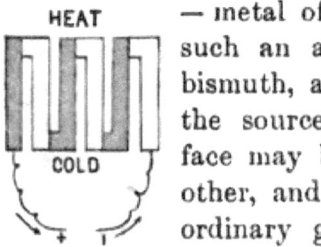

Fig. 197.

Such contrivances for generating electric currents are called *thermo-electric batteries*. They are seldom used, inasmuch as the best of them transform less than one per cent of the heat energy given out by the source of heat.

If the source of heat is feeble or distant, the feeble current may serve to measure the difference of temperature between the ends of the bars turned toward the heat and the other ends, which are at the temperature of the air. The apparatus, when used for this purpose, is called a *thermo-pile*, or a *thermo-multiplier*. A combination (Fig. 198) of as many as thirty-six pairs of antimony and bismuth bars, connected with a very sensitive galvanometer, constitutes an exceedingly delicate *thermoscope* and *thermometer*. Changes of temperature that would not produce a perceptible change in an ordinary thermometer, can, by the use of a thermo-electric pile, be made to produce large deflections of the galvanometer needle. Heat radiated from the body of an insect several inches from the pile may cause a sensible deflection.

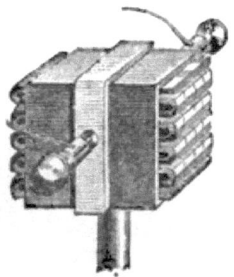

Fig. 198.

Section XX.

STATIC ELECTRICITY.

200. Mechanical Energy transformed into Electric Potential Energy or Electrification.

Experiment 172. — Prepare an insulated stool (Fig. 199) by placing a square board on four *dry* and *clean* glass tumblers, used as legs. Let a person, whom we will call John, stand on this stool, and let a second person, James, strike John a few times with a cat's fur. Then let James bring the knuckle of a finger near to some part of John's person, for instance his hand, chin, or nose; an electric spark will pass between the two, and both will experience a slight shock. The length of the spark shows that the electricity is urged by a high E.M.F., like the induced currents of the magneto-machine and induction coil.

As mechanical energy is transformed into potential energy in the act of bending a bow or stretching a rubber band, in other words, a peculiar molecular stress is developed thereby, so by the expenditure of mechanical energy in separating the fur from the boy at the end of each stroke there is developed a phase of potential energy; the bodies in which it is developed are said to be in a state of *electrification*, in other words, there exists between them a form of *electric stress*. The electrified bodies are sometimes said to be "charged" with electricity.

Fig. 199.

201. Electroscope.

Experiment 173. — Suspend in a loop, tied in a white silk thread, a strip of "Dutch metal," so that the two vertical portions may be near each other. After John has been struck a few times with the fur, let him bring a finger gradually near the upper extremity of the foil; the two portions of the foil gradually diverge, as in Figure 200, indicating the action of an unusual force between them.

Fig. 200.

Any arrangement, like that of the foil just described, intended to detect the presence of electrification, is called an *electroscope*. One of the most common and useful electroscopes consists of one or two pith-balls, made from the pith of elder or sunflower, suspended by silk thread. If an electroscope is brought near to either pole of a secondary wire of an induction coil, a similar electrification is manifested by the poles. Likewise, by means of very delicate electroscopes, the poles of a galvanic battery, or of a thermo-battery, are found to be feebly electrified.

202. Attractions and Repulsions.

Experiment 174. — Poise a flat wooden ruler on an inverted bottle or flask, having a round bottom, as in Figure 201. Draw a rubber comb two or three times through your hair, or rub it with a woollen cloth, and place it near one end of the ruler; instantly the ruler moves toward the comb.

Experiment 175. — Hold the comb over a handful of bits of tissue paper; the papers quickly jump to the comb, stick to it for an instant, and then leap energetically from the comb. The papers are first attracted to the comb, but in a short time acquire some of its electrification, and then are repelled.

Fig. 201.

Experiment 176. — Support a plate of window glass (Fig. 202) about two inches from a table. Rub its upper surface with a silk handkerchief, and place pith-balls or bits of tissue paper on the table beneath the glass. They will dance up and down between the plate and table in a lively manner.

Fig. 202.

203. Two States of Electricity. — It is quite apparent that we are now dealing with a very different class of

electrical phenomena from any that we have previously observed. It is also quite as obvious that we are dealing with electricity in a very different state or condition from that in which we have before studied it. Hitherto we have studied only those phenomena produced by electricity when in motion; and, inasmuch as when in that state its energy is expended in work, or transformed into some other form of energy as rapidly as it is generated, there was no such thing as an *accumulation* of electricity. In our late experiments there is wanting anything like a current; but, on the other hand, we find that electricity in this new state may accumulate, be stored up, and remain in a quiescent state for an indefinite time. In the latter state it is incapable of affecting a magnetic needle, magnetizing, generating heat, illuminating, producing decomposition, or giving shocks. But in this state of apparent repose it may attract and afterwards repel light bodies in the vicinity of the body in which it resides. These attractions and repulsions are quite distinct from the attractions and repulsions which occur between parallel currents.

This state of electricity is called *static*, in distinction from the current state, which is often called *dynamic*. We have seen that, under certain conditions, electricity may change from one state to the other, as when the electricity which had accumulated in the boy on the insulated stool passed to the other boy, producing, in its current state, both illuminating and physiological effects; and again, when a circuit is broken, the current ceases, but electricity accumulates in the wire. We have also learned that electricity of high E.M.F., such as is most readily developed by friction, exhibits the static phenomena, *i.e.* attractions and repulsions, most strikingly.

204. Two Kinds of Electrification.

Experiment 177. — Bend a small glass tube into the form represented by A (Fig. 203), insert one end in a block of wood B for a base; and suspend from the tube a pith-ball C by a silk thread. Rub a glass rod D with a silk handkerchief, and present it to the ball; attraction at first occurs, followed by repulsion after contact. Now rub a stick of sealing-wax, or a hard-rubber ruler, with flannel, and present it to the ball, which is in a condition such that it is repelled by the electrified glass; it is attracted by the electrified sealing-wax. We are led to suspect that the sealing-wax possesses a different kind of electrification from that of the glass. Let us further test the matter.

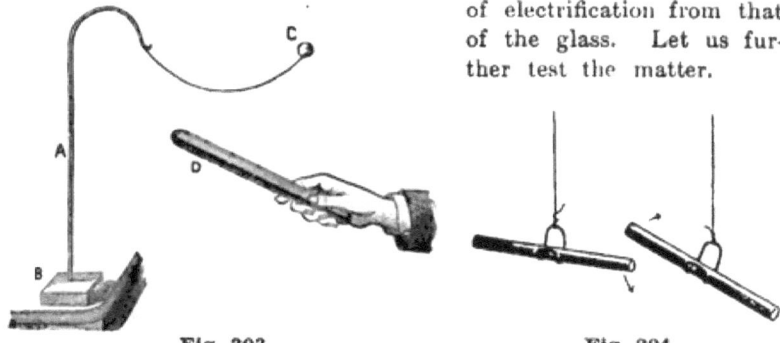

Fig. 203.   Fig. 204.

Experiment 178. — Suspend two glass rods that have each been rubbed with silk in two wire stirrups (Fig. 204), and present them to each other; they repel one another. Suspend two sticks of sealing-wax that have been rubbed with flannel in the same manner; the same result follows. Now, in a like manner, present one of the glass rods and one of the sticks of sealing-wax to each other; they attract one another.

It is evident (1) that there are *two kinds or conditions of electrification*, or, for convenience, we sometimes say *two kinds of electricity*; (2) that they are so related to each other that *like kinds repel, and unlike kinds attract each other*. The two kinds are usually distinguished from each other by the names *positive* and *negative*, or, more briefly, as $+E$ and $-E$. The former is, by definition,

230   ELECTRICITY AND MAGNETISM.

such as is developed on glass when rubbed with silk, and the latter is the kind developed on sealing-wax when rubbed with flannel. There is no reason, except custom, for calling the one positive rather than the other.

Experiment 179.— Once more electrify a stick of sealing-wax with a flannel, and present it to a pith-ball, and after the ball is repelled, bring the surface of the flannel which had electrified the rod near the ball; the ball is attracted by it, showing that the rubber is also electrified and with the opposite kind to that which the sealing-wax possesses.

One kind of electrification is never developed alone; when two bodies are rubbed together they become equally but oppositely electrified.

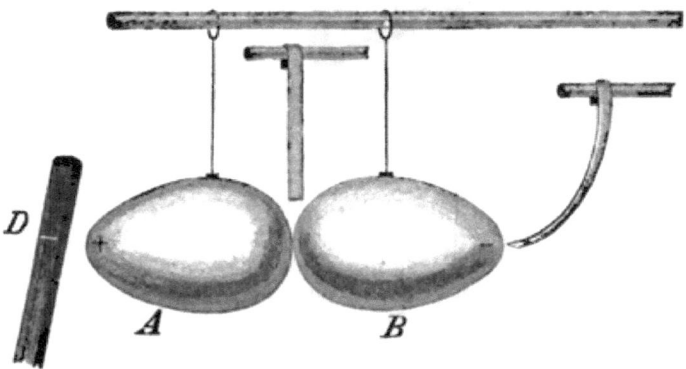

Fig. 205.

205. Induction.

Experiment 180.— Suspend by silk threads from a glass tube two egg-shells covered with tin foil, so as to touch each other, as in Figure 205. Bring near to one end of the shells, but not to touch, a sealing-wax rod excited with flannel, and therefore having — E. While the rod is in this position, carry a thin strip of tissue paper, or a pith-ball suspended by a silk thread, along the eggs. The paper is attracted most strongly at the ends; but in the middle, where the shells are in contact, there is very little electrification. Separate B from A about

10^{cm}, while the rod D is still in position. Then place D midway between A and B; the rod repels B and attracts A. It appears that when the two shells touched each other, thereby constituting practically one body, that the shells were oppositely electrified, as represented by the signs + and − in the diagram; and when the two bodies were separated, they retained their opposite charges.

We learn from this experiment that by *induction* we may charge at the same time two bodies, one with $+ E$ and the other with $- E$.

206. Discharge.

Experiment 181. — Bring the two shells oppositely charged near each other; when near enough they exhibit mutual attraction for each other. On bringing them still nearer, a spark passes between them, their mutual attraction suddenly ceases, and on testing them with an electroscope, it is found that both have lost their electrification, *i.e.* both have become *discharged*.

When two bodies equally and oppositely electrified are brought together, both become discharged. During the process of discharge, the electricity which was previously in a condition of rest, or a static state, assumes a condition of motion, or a dynamic state, as is shown by a spark passing between the two bodies when brought near each other. One of the bodies (that positively charged) is at a potential higher than that of the earth, the other is at a lower potential. When they are brought sufficiently near, the tendency of the electricity to pass from the region of higher potential becomes strong enough to penetrate the insulating air and establish a condition of equilibrium. In this particular case the result is zero potential or no electrification; but in general both bodies would be left at a like condition of electrification, its sign depending upon the sign of that electricity which was in excess.

We may now understand how it is that an electrified body attracts to itself light bodies in its vicinity. For example, a stick of sealing-wax, excited with $- E$, brought near a pith-ball, induces $+ E$ next itself, and repels $- E$ to its farthest side; then, of course, attraction follows. There is the same attraction between heavy bodies, but usually not sufficient to produce motion.

207. Insulation. — A body that is to receive a permanent charge of electricity must be insulated, *i.e.* have no connection with the earth through a conducting substance. Some of the best insulating substances are *dry air, ebonite, shellac, resins, glass, silks,* and *furs.* Moisture injures the insulation of bodies; hence experiments succeed best on dry, cold days of winter, when moisture of the air is least liable to be condensed on the surfaces of apparatus, *especially if they are kept warm.*

Section XXI.

ELECTRICAL MACHINES. — CONDENSERS, ETC.

208. Plate Machine. — An electrical machine is an instrument intended for transforming mechanical energy into the energy of electrification. The *plate machine* (Fig. 206) consists of a *conductor* A, a glass plate B, a rubber C made of two cushions covered with a preparation which facilitates the excitation, and a brass chain E used to connect the cushions with the earth. An extension of the conductor consists of a comb D whose pointed teeth are turned towards the plate. When the plate is turned in the direction indicated by the arrow, it passes between the rubbers, and the friction causes $+$ E to collect on the plate and $-$ E on the rubber. The electrified portion of the plate then comes opposite the comb, when it polarizes the conductor, attracting $-$ E and repelling $+$ E. But the E escapes from the points of the comb to the plate, neutralizes the $+E$ of the plate, and thereby leaves the conductor charged with $+E$.

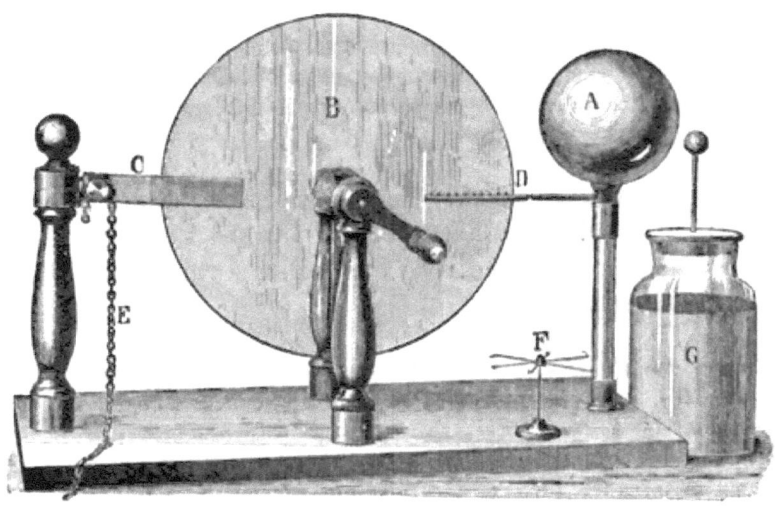

Fig. 206.

209. Electrophorus. — This apparatus is used to incite electrification by induction. It consists of a shallow iron dish A (Fig. 207) filled with sealing-wax. At the center of the dish is a protuberance B which extends just through the wax. A flat brass disk C has a glass insulating handle.

Experiment 182. — Strike the surface of the wax a few times with a cat's fur, or rub it with a dry flannel. The wax becomes electrified with $-E$. Place the disk C upon it. The $+E$ of the disk is *bound* (*i.e.* held by the attraction of) the $-E$ of the wax, but the $-E$ of the disk is repelled by the $-E$ of the wax and passes through the protuberance B to the dish below, and thence to the earth. Consequently when the disk C is raised by

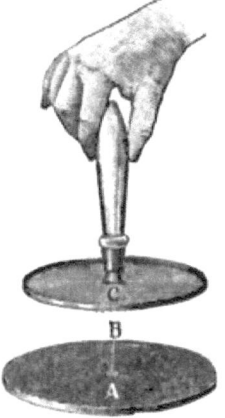

Fig. 207.

the insulating handle from the wax, it is charged with $+$ E, and the charge can be transferred to any body (*e.g.* a Leyden jar), and then the disk can be recharged by replacing it on the wax. This may be repeated many times without sensibly reducing the inductive power of the wax.

The Holtz machine (Fig. 139) is a sort of a continuous electrophorus which is capable of developing electrification by induction so rapidly and continuously as to give an almost incessant flow of sparks between the two conductors.

210. Condenser. — A very important adjunct to an electrical machine is a *condenser* of some kind, by means of which a large quantity of electricity can be collected on a small surface.

Experiment 183. — Let a person stand on an insulated stool (page 226), and place one hand on the conductor of a machine. Let the other open hand press against a plate of glass or a disk of vulcanite, held on the open hand of a second person standing on the floor. After a few turns of the machine, let the hand that has been on the prime conductor grasp the free hand of the second person. Quite a shock will be felt by both.

It is evident that by this process an unusual quantity of electricity had collected previous to the discharge. The explanation is simple. The hand of the first person, charged with $+$ E, acts by induction through the glass upon the second person, attracting $-$ E to the surface of the glass with which his hand is in contact, and repelling $+$ E to the earth. Thus, through their mutual attraction, the two kinds of electricity become, as it were, heaped up opposite each other, and yet are prevented, by the insulating glass, from uniting.

211. Leyden Jar. — The most convenient form of condenser is the *Leyden jar* (Fig. 208). It is a wide-mouthed glass jar lined on the inside and outside for about two-thirds its height with tin foil. Through the stopper passes a brass rod terminating at its upper extremity in a brass ball a, and at the other extremity in a brass chain which touches the inner coating of tin foil.

The jar may be charged by connecting one of its coatings with the conductor of an electrical machine, and the other with the earth. Or it may be charged by connecting the outside coating with one of the poles of the Holtz machine, and bringing the other pole near to the ball leading from the inner coating. To discharge the jar, connect the outer coating with the knob of the jar. To avoid a shock in so doing a *discharger* is used (Fig. 209), which consists of a bent wire terminating at each end with metal balls. The wire is held by a glass insulating handle.

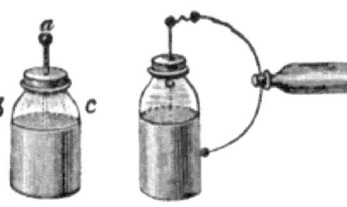

Fig. 208.   Fig. 209.

212. Electrification confined to the External Surface.
Experiment 184. — Place a tin fruit-can on a clean, dry glass tumbler (Fig. 210). Fasten a circular disk a of tin 15^{mm} in diameter to one end of a rod of sealing-wax. Charge the can heavily with electricity from an electrical machine. Through an orifice c in the can introduce the disk, and touch the interior surface of the can. Withdraw the disk, and present it to an electroscope. It shows no electrification. Now touch the exterior surface of the can with the disk and present it to the electroscope; it is found to be electrified.

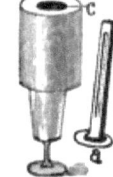

Fig. 210.

This experiment shows that no electricity can be found inside of a hollow charged conductor; or, roughly stated, *a static charge of electricity resides on the exterior surface of a conductor.*

213. Effect of Points. — An *electrical flyer*, F (Fig. 206), consists of a cap of metal resting upon a pointed wire, which serves as a pivot. The cap has pointed wires branching out from it like the spokes of a wheel, bent near the ends and turned in the same direction. If this is placed on an insulating stand and connected with the conductor of an electrical machine when in operation, the air particles around the electrifying points become excited like so many pith-balls, and are rapidly repelled, producing a continuous current of air issuing from the points. The reaction of these air-particles causes the wheel to revolve in the opposite direction. As we might reasonably expect, currents of excited air-particles issuing from the points on an excited conductor serve to carry away with them portions of the charge, so that *the effect of points on an electrified insulated body is greatly to facilitate its discharge.*

214. Lightning. — Certain clouds which have formed very rapidly are highly charged with electricity, usually positively charged. The surface of the earth and objects thereon immediately beneath the cloud are charged inductively with the opposite kind of electricity. The cloud and the earth correspond to the coatings, and the intervening air to the glass of a huge Leyden jar. The charge in the earth and that in the cloud hold each other prisoner by their mutual attraction, until, as the charges accumulate, the attraction becomes great enough to disrupt the insulating medium, *i.e.* the intervening air, when a discharge takes place. It is the accumulation of induced electricity on elevated objects, such as buildings and trees, that offers an attraction for the opposite electricity of the cloud, and renders them especially liable to be struck by lightning.

215. Lightning-Rods. — The flash will pass along the line of least resistance. A *good* lightning conductor offers a peaceful means of communication between the earth and a cloud; it leads the electricity of the earth gently up toward the cloud, and allows it to combine with its opposite without disturbance, thereby so far discharging the cloud as to prevent a lightning stroke; or, if the tension is too great to be thus quietly disposed of, the flash strikes downward, and is led harmlessly to the earth by the conductor. *An ill-constructed lightning-rod may be worse than none.* A rod should be made of good conducting material, so large that it will not be melted, and free from loose joints. The lower end should be buried in earth that is always moist, and the upper end should terminate in several sharp points.

CHAPTER VII.

SOUND.

Section I.

STUDY OF VIBRATIONS AND WAVES.

The subjects of Sound-waves and Light-waves, which we are about to study, have two important characteristics in common that distinguish them from the subjects already studied. First, each of them affects its peculiar organ of sense, the ear or the eye. Secondly, both originate in vibrating bodies, and reach us only by the intervention of some medium capable of being set in vibration.

216. Period of Vibration.

Experiment 185. — Suspend an iron ball by a string, as in Experiment 71, cause it to vibrate, and, watch in hand, ascertain the number of vibrations made in a given number of seconds; *e.g.* 60 seconds. Then, remembering that all the vibrations are made in equal intervals of time, ascertain the *period of vibration* of this pendulum; *i.e.* the time it takes to make each vibration, using the formula

$$t = \frac{s}{n},$$

in which $t =$ the period, and $n =$ the number of vibrations made in s seconds.

217. Direction of Vibration.

Experiment 186. — Grasp one end of a small rod or yardstick in a vice, pull the free end one side, and set it in vibration. Pluck a string of a piano or violin. Note that the motions of all the bodies which thus far we have caused to vibrate are at *right angles to their length*. These are called *transverse vibrations*.

Experiment 187. — Hang up a spiral spring or elastic cord with a small weight attached at the lower end; lift the weight, and, dropping it, notice that the cord vibrates *lengthwise*. This is a case of *longitudinal vibration*. There may also be *torsional vibrations*, for example children often amuse themselves by producing these by twisting a window cord and tassel.

218. Propagation of Vibration; Waves.

Experiment 188. — Take a rubber cord about the size of an ordinary lead-pencil and 12 feet long. Attach at intervals a few glass beads and fasten one end of the cord to the wall of the room. Hold the free end in the hand and draw the cord out so as to be nearly horizontal. By quick movements of the hand in a horizontal or a vertical direction set this end in vibration. Notice that these vibrations are communicated from point to point along the cord, and that each point in the cord successively goes through a vibration precisely similar to that held in the hand. Fix the eyes upon any one of the beads; it simply vibrates transversely. Observe the cord as a whole; *waves* traverse it from end to end, but it is easy to see that it is only a *form* that traverses it; the beads and all other points of the cord move transversely. These successive transverse movements give rise to the *wave-line* into which the cord is thrown.

219. Wave-Length and Amplitude.

— Imagine an instantaneous photograph taken of the cord along which continuous waves are passing. It would appear much like the curved line CD (Fig. 211). This curved line represents what is known as a simple *wave-line*.

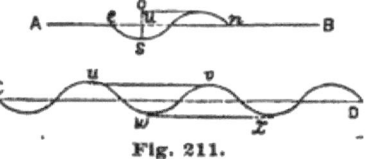

Fig. 211.

The distance from any vibrating point to the nearest point which is in exactly the same stage of its vibration is called a *wave-length*, as *wx*, *uv*, or *en*.

The distance between the extreme positions of a vibrating point or the length of its journey is called the *amplitude of the wave* or the *amplitude of vibration*.

220. Reflection of Waves; Interference.

Experiment 189. — Stretch the cord horizontally between two elevated points, and pluck it with the hand or strike it with a stick near one end, and send along it a single pulse, forming a crest on the rope (A, Fig. 212). This travels to the other end, and there we see it reflected and inverted (B).

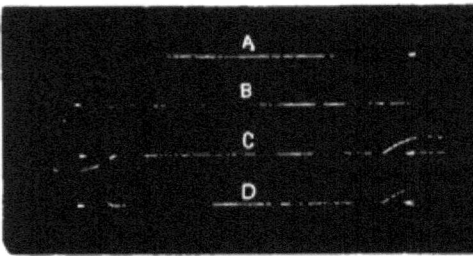

Fig. 212.

Experiment 190. — Just at the instant of reflection, start a second crest; these two, the crest and the returning inverted crest or trough (C), are now travelling along the rope in opposite directions, and must meet at some point. This point will be urged upward by the crest and downward by the trough, and so its motion will be due to the difference of the two forces.

Experiment 191. — Send along the rope, first a trough, then a crest; now two crests (D) will meet near the middle of the rope, and the motion here will be due to two forces acting in the same direction, so that the resulting crest will be greater than either of the original ones.

This action on a single point of two pulses, or two trains of waves, no matter if from different sources, is termed *interference*. The resulting motion may be greater or less than that due to either pulse alone, or it may be zero.

221. Stationary Vibrations, Nodes, etc.

Experiment 192. — Hold one end of the cord while the other is fixed, and send along it a regular succession of equal pulses from the

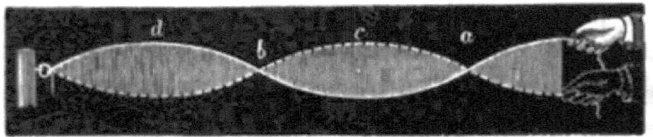

Fig. 213.

vibrating hand; it will be easy, by varying the tension and rate a little, to obtain a succession of hazy spindles (Fig. 213), separated by points that are nearly or quite at rest. Unlike the earlier experiments, the waves here do not appear to travel along the tube; yet in reality they do traverse it. The deception is caused by stationary points being produced by the interference of the advancing and retreating waves.

This interference of direct and reflected waves gives rise to the important class of so-called *stationary vibrations*. The points of least motion, as *a* and *b*, are called *nodes;* the points of greatest motion, *c* and *d*, are called *antinodes;* and the portion of the rope between two nodes, as *ab*, is a *ventral segment*.

222. Longitudinal Waves.

Experiment 193. — Figure 214 represents a brass wire wound in the form of a spiral spring, about 12 feet long. Attach one end to a cigar-box, and fasten the box to a table. Hold the other end H of the spiral firmly in one hand, and with the other hand insert a knife-blade between the turns of the wire, and quickly rake it for a short distance along the spiral toward the box, thereby crowding closer together for a little distance (B) the turns of wire in front of the hand, and leaving the turns behind pulled wider apart (A) for about an equal distance. The

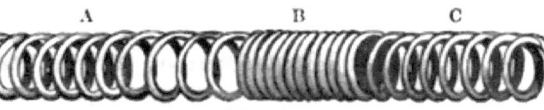

Fig. 214.

crowded part of the spiral may be called a *condensation*, and the stretched part a *rarefaction*. The condensation, followed by the rarefaction, runs with great velocity through the spiral, strikes the box, producing a sharp thump; is reflected from the box to the hand, and from the hand again to the box, producing a second thump; and by skilful manipulation three or four thumps will be produced in rapid succession. If a piece of twine be tied to some turn of the wire, it will be seen, as each wave passes it, to receive a slight jerking movement forward and backward in the direction of the length of the spiral.

How is energy transmitted through the spring so as to deliver the blow on the box? Certainly not by a bodily movement of the spiral as a whole, as might be the case if it were a rigid rod. The movement of the twine shows that the only motion which the coil undergoes is a vibratory movement of its turns. Here, as in the case of water-waves, energy is transmitted through a medium by the transmission of vibrations.

There are two important distinctions between these waves and those which we have previously studied: the former consist of condensations and rarefactions; the latter, of elevations and depressions. In the former, the vibration of the parts is in the same line with the path of the wave, and hence these are called *longitudinal waves;* in the latter, the vibration is across its path; they are therefore called *transverse waves.*

A wave cannot be transmitted through an inelastic soft iron spiral. *Elasticity is essential in a medium, that it may transmit waves composed of condensations and rarefactions; and the greater the elasticity, the greater the facility and rapidity with which a medium transmits waves.*

223. Air as a Medium of Wave-Motion. — May not air and other gases, which are elastic, serve as media for waves?

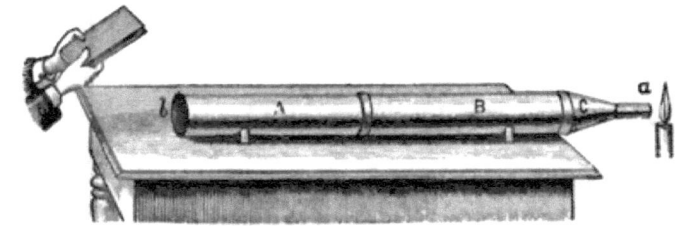

Fig. 215.

Experiment 194. — Place a candle flame at the orifice *a* of the tube (Fig. 215), and strike the table a sharp blow with a book near the orifice *b*. Instantly the candle flame is quenched. The body of air in the tube serves as a medium for transmission of motion to the candle.

Was it the motion of a current of air through the tube, a miniature wind, or was it the transfer of a vibratory motion? Burn touch-paper[1] at the orifice *b*, so as to fill this end of the tube with smoke, and repeat the last experiment.

Evidently, if the body of the air is moved along through the tube, the smoke will be carried along with it. The candle is blown out as before, but no smoke issues from the orifice *a*. It is clear that there is no translation of material particles from one end to the other, — nothing like the flight of a rifle bullet. The candle flame was struck by something like a *pulse* of air, not by a *wind*.

224. How a Wave is propagated through a Medium. — The effect of applying force with the hand to the spiral spring is to produce in a certain section (B, Fig. 214) of the spiral a crowding together of the turns of wire, and at A a separation; but the elasticity of the spiral instantly causes B to expand, the effect of which is to produce a crowding together of the turns of wire in front of it, in the section C, and thus a forward movement of the condensation is made. At the same time, the expansion of B causes a filling up of the rarefaction at A, so that this section is restored to its normal state. This is not all: the folds in the section B do not stop in their swing when they have recovered their original position, but, like a pendulum, swing beyond the position of rest, thus producing a rarefaction at B, where immediately before there was a condensation. Thus a forward movement of the rarefaction is made, and thus a pulse or wave is transmitted with uniform velocity through a spiral spring, air, or any *elastic* medium.

225. Graphical Method of Studying Vibrations.
Experiment 195. — Attach, by means of sealing-wax, a bristle or a fine wire to the end of one of the prongs of a large steel fork (like

---

[1] To prepare touch-paper, dissolve about a teaspoonful of saltpetre in a half-teacupful of hot water, dip unsized paper in the solution, and then allow it to dry. The paper produces much smoke in burning, but no flame.

a tuning-fork, but larger) called a *diapason*. Set the fork in vibration, and quickly draw the point of the bristle lightly over a smoked glass (A, Fig. 216). A beautiful wavy line will be traced on the glass, each wave corresponding to a vibration of the prong when vibrating as a whole.

Fig. 216.

Next, tap the fork, near its stem, on the edge of a table, and trace its vibrations on a smoked glass as before. You will generate a similar set of waves, but, running over these, is another set, of much shorter period, like No. 3 of Figure 230, showing that the prong vibrates, not only as a whole, but in parts. The serrated wavy line produced represents the resultant of the combined vibrations, and may be called a *complex wave-line*.

QUESTIONS.

1. In what kind of motion does all wave-motion originate?
2. Watch the waves of the ocean moving landward; what is it that advances?
3. Throw a cord into wavy motion by the movement of your hand; upon what do the number and the length of the waves which traverse the cord at any given time depend?
4. How is a node produced?
5. How do the vibrations in longitudinal waves differ from the vibrations in transverse waves?
6. Are the vibrations in air-waves longitudinal or transverse?

Section II.

SOUND-WAVES.

226. How Sound-waves Originate. — Listen to a sounding church-bell. It produces a sensation; it is heard. The ear is the organ through which the sensation of hearing is produced. The bell is at such a distance that it cannot act directly on the ear; yet something must act on the ear, and it must be the bell which causes that something to act.

Commencing at the origin of sound, let the first inquiry be, How does a sounding body differ from a silent body?

Experiment 196. — Strike a bell or a glass bell-jar, and touch the edge with a small cork ball suspended by a thread; you not only *hear* the sound, but, at the same time, you *see* a tremulous motion of the ball, caused by a motion of the bell. Touch the bell gently with a finger, and you feel a tremulous motion. Press the hand against the bell; you stop its vibratory motion, and at that instant the sound ceases. Strike the prongs of a tuning-fork, press the stem against a table: you hear a sound. Thrust the ends of the prongs just beneath the surface of water; the water is thrown off in a fine spray on either side of the vibrating fork. Watch the strings of a piano, guitar, or violin, or the tongue of a jews-harp, when sounding. You can *see* that they are in motion.

Sound-waves originate in a vibrating body.

227. How Sound-waves Travel. — How can a bell, sounding at a distance, affect the ear? If the bell while sounding possesses no peculiar property except motion, then it has nothing to communicate to the ear but motion. But motion can be communicated by one body to another at a distance only through some medium.

Do sound-waves require a medium for their communication?

Experiment 197. — Lay a thick tuft of cotton-wool on the plate of an air-pump, and on this, face downward, place a loud-ticking watch, and cover with the receiver. Notice that the receiver, interposed between the watch and your ear, greatly diminishes the sound, or interferes with the passage of *something* to the ear. Take a few strokes of the pump and listen; the sound is more feeble, and continues to grow less and less distinct as the exhaustion progresses, until either no sound can be heard when the ear is placed close to the receiver, or an extremely faint one, as if coming from a great distance. The removal of air from a portion of the space between the watch and your ear destroys the sound. Let in the air again, and the sound is restored.

Sound-waves cannot travel through a vacuum, i.e. without a medium.

Boys often amuse themselves by inflating paper bags, and with a quick blow bursting them, producing with each a single loud report. First the air is suddenly and greatly condensed by the blow, and the bag is burst; the air now, as suddenly and with equal force, expands, and by its expansion condenses the air for a certain distance all around it, leaving a rarefaction where just before had been a condensation. If many bags were burst at the same spot in rapid succession, the result would be that alternating shells of condensation and rarefaction would be thrown off, all having a common center, enlarging as they advance, like the waves formed by stones dropped into water; except that, in this case, the waves are not like rings, but hollow globes; not circular, but spherical. In this manner sound-waves produced by the vibration of a sounding body travel through the air.

As a wave advances, each individual air-particle concerned in its transmission performs a short excursion to and fro in the direction of a straight line radiating from the center of the shells or hollow globes. *A sound-wave travels its own length in the time that a particle occupies in going through one complete vibration so as to be ready to start again.*

Experiment 198. — Take a strip of black cardboard 4.5 inches × 1 inch. Cut a slit about one-sixteenth of an inch wide lengthwise and centrally through the strip nearly, from end to end. Place the slit over the dotted line at the bottom of Figure 217, and draw the book along underneath in the direction of the arrow. Imagine that the short white dashes seen through the slit represent a series of air-particles, and the slit itself represents the direction in which a series of sound-waves are travelling. It will be seen that each air-particle moves a little to and fro in the direction in which the sound travels and comes back to its starting-point; but the condensations and rarefactions, represented by a group (half a wave-length) of dots being alternately closer together or farther apart, are transmitted through the whole series of air-particles.

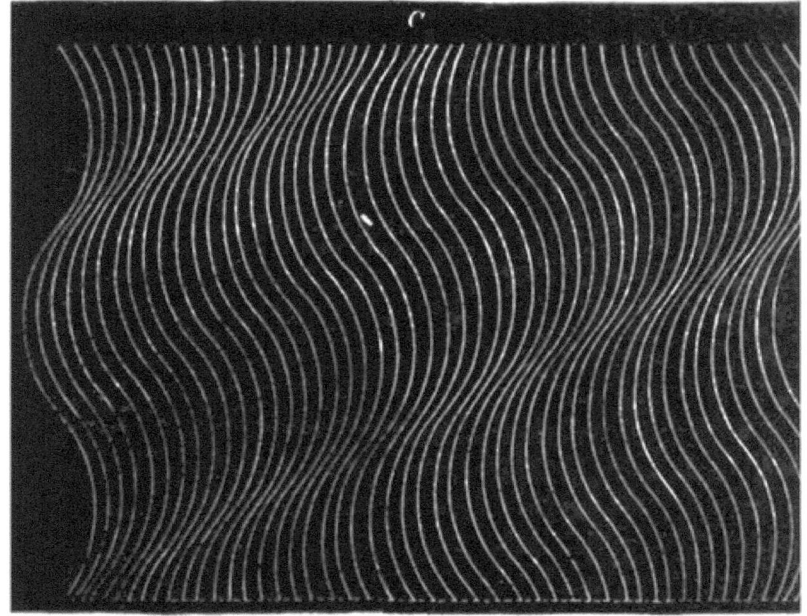

Fig. 217.

228. What Sound Is. — *Sound is a sensation caused usually by waves of air beating upon the organ of hearing.*

229. Solids and Liquids as Media for transmitting Sound-waves.

Experiment 199. — Lay a watch, with its back downward, on a

long board (or table), near to one of its ends, and cover the watch with loose folds of cloth till its ticking cannot be heard through the air in any direction at a distance equal to the length of the board. Now place the ear in contact with the farther end of the board, and you will hear the ticking of the watch very distinctly.

Experiment 200. — Place one end of a long pole on a cigar box, and apply the stem of a vibrating diapason to the other end; the sound-vibrations will be transmitted through the pole to the box, and a loud sound will be given out by the box, as though that, and not the tuning-fork, were the origin of the sound.

Experiment 201. — Place the ear to the earth, and listen to the rumbling of a distant carriage; or put the ear to one end of a long stick of timber, and let some one gently scratch the other end with a pin.

Solids and liquids, as well as gases, transmit sound-vibrations.

Section III.

VELOCITY OF SOUND-WAVES.

230. The Velocity of Sound-waves depends on the Elasticity and Density of the Medium. — The relation of velocity to the density and elasticity of gases, as ascertained by careful experiment, is as follows: *the velocity of sound-waves in gases is directly proportional to the square root of their elasticity, and inversely proportional to the square root of their respective densities.*

The velocity of sound-waves in air at 0° C. is (333^m) 1093 feet per second. The velocity increases nearly two feet for each degree centigrade. At the temperature of 16° C. (60° F.) we may reckon the velocity of sound-waves at about (342^m) 1125 feet per second.

The greater density of solids and liquids, as compared with gases, tends, of course, to diminish the velocity of sound-waves; but their greater incompressibility more than compensates for the decrease of velocity occasioned by the increase of density. As a general rule, solids are more incompressible than liquids; hence, sound-waves generally travel faster in the former than in the latter. For example, sound-waves travel in water about 4 times as fast as in air, and in iron and glass 16 times as fast.

Section IV.

REFLECTION OF SOUND-WAVES. — ECHOES.

231. Reflection. — In the experiment with the spiral spring, waves were reflected from the box to the hand, and from the hand to the box. When a sound-wave meets an obstacle in its course, it is reflected; and the sound resulting from the reflected waves is often called an *echo*, or, when they are many times reflected so that the sound becomes nearly continuous, a *reverberation*.

232. Sound-waves reflected by Concave Mirrors.

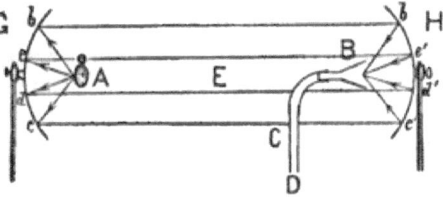

Fig. 218.

Experiment 202. — Place a watch at the focus A (Fig. 218) of a concave mirror G. At the focus B of another concave mirror H, place the large opening of a small tunnel, and with a rubber connector attach the bent glass tube C to the nose of the tunnel. The extremity D being placed in the ear, the ticking of the watch can be heard very distinctly, as though it were somewhere near the mirror H. Though the mirrors be 12 feet apart, the sound will be louder at B than at an intermediate point E.

How is this explained? Every air-particle in a certain radial line, as Ac, receives and transmits motion in the direction of this line; the last particle strikes the mirror at c, and being perfectly elastic, bounds off in the direction cc', communicating its motion to the particles in this line. At c' a similar reflection gives motion to the air particles in the line c'B. In consequence of these two reflections, all divergent lines of motion as Ad, Ae, etc., that meet the mirror G, are there rendered parallel, and afterwards rendered convergent at the mirror H. The practical result of the concentration of this scattering energy is, that a sound of great intensity is heard at B. The points A and B are called the *foci* of the mirrors. The front of the wave as it leaves A is convex, in passing from G to H it is plane, and from H to B concave. If you fill a large circular tin basin with water, and strike one edge with a knuckle, circular waves with concave fronts will close in on the centre, heaping up the water at that point.

Long "whispering-galleries" have been constructed on this principle. Persons stationed at the foci of the concave ends of the long gallery can carry on a conversation in a whisper which persons between cannot hear.

The external ear is a wave-condenser. The hand held concave behind the ear, by its increased surface, adds to its efficiency. An ear-trumpet, by successive reflections, serves to concentrate, at the small orifice opening into the ear, the sound-waves that enter at the large end.

Section V.

INTENSITY OF SOUND.

233. Intensity depends on the Amplitude of Vibration. — Gently tap the prongs of a tuning-fork and dip them into water, — the water is scarcely moved by them; increase the force of the blow, — the vibrations become wider, and the water spray is thrown with greater force and to a greater distance. The same thing occurs when the fork vibrates in the air; though we do not see the air-particles as they are batted by the moving fork, yet we feel the effects as a sound sensation, and we judge of their energy by the intensity of the sensation which they produce. *Loudness* of sound refers to the intensity of a sensation. We have no standard of measurement for a sensation, so we are compelled to measure the intensity of the sound-wave, knowing at the same time that *loudness is not proportional to this intensity*. Unfortunately, the expressions *loudness* and *intensity of sound-wave* are often interchanged. The intensity of a vibration is measured by the energy of the vibrating particle. It is clear that if the amplitude of vibration of a particle is doubled while its period remains constant, its velocity is doubled, and its energy is increased fourfold. Hence, (1) *measured mechanically, the intensity of a sound-wave is proportional to the square of the amplitude of the vibrations of the vibrating body.*

234. Intensity depends upon the Density of the Medium. — In the experiment with the watch under the receiver of the air-pump (page 245), the sound grew feebler as the air became rarer. Aëronauts are obliged to exert themselves more to make their conversation heard when they reach great hights than when in the denser lower air. (2) *The intensity of sound-waves increases with the density of the medium in which they are produced.*

235. Intensity depends on Distance. — It is a matter of every-day observation that the loudness of a sound diminishes very rapidly as the distance from the source of the waves to the ear increases. As a sound-wave advances in an ever-widening sphere, a given amount of energy becomes distributed over an ever-increasing surface; and as a greater number of particles partake of the motion, the individual particles receive proportionately less energy; hence it follows, — as a consequence of the geometrical truth, that "the surface of a sphere varies as the square of its radius," — that (3) *the intensity of a sound-wave varies inversely as the square of the distance from its source.* For example, if two persons, A and B, are respectively 500 and 1000 rods from a gun when it is discharged, the waves that reach A will be four times as intense as the same when they reach B.

236. Speaking-Tubes.

Experiment 203. — Place a watch at one end of the long tin tube (Fig. 215), and the ear at the other end. The ticking sounds very loud, as though the watch were close to the ear.

Long tin tubes, called *speaking-tubes*, passing through many apartments in a building, enable persons at the distant extremities to carry on conversation in a low tone of voice, while persons in the various rooms through which the tube passes hear nothing. The reason is that the sound-waves which enter the tube are prevented from expanding, consequently the intensity of sound is not affected by distance, except as its energy is wasted by friction of the air against the sides of the tube.

Section VI.

REËNFORCEMENT OF SOUND-WAVES AND INTERFERENCE OF SOUND-WAVES.

237. Reënforcement of Sound-waves.

Experiment 204. — Set a diapason in vibration; you can scarcely hear the sound unless it is held near the ear. Press the stem against a table; the sound rings out loud, but the waves seem to proceed from the table.

When only the fork vibrates, the prongs presenting little surface cut their way through the air, producing very slight condensations, and consequently waves of little intensity. When the fork rests upon the table, the vibrations are communicated to the table; the table with its larger surface throws a larger mass of air into vibration, and thus greatly intensifies the sound-waves. The strings of the piano, guitar, and violin owe as much of their loudness of sound to their elastic sounding-boards, as the fork does to the table.

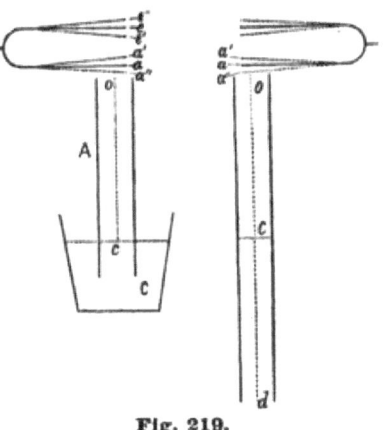

Fig. 219.

238. Reënforcement by Bodies of Air; Resonators.

Experiment 205. — Take a glass tube A (Fig. 219), 16 inches long and 2 inches in diameter; thrust one end into a vessel of water C, and hold over the other end a vibrating diapason B that makes (say) 256 vibrations in a second. Gradually lower the tube into the water, and when it reaches a certain depth, *i.e.* when the column of air *oc* attains a certain length, the sound of the fork becomes very loud; continuing to lower the tube, the sound rapidly dies away.

Columns of air are thus found to serve, as well as sounding-boards, to *reënforce* sound-waves. The instruments which enclose the columns of air are called *resonators*. Unlike sounding-boards, they can respond loudly to only one tone, or to a few tones of widely different pitch.

How is this reënforcement effected? When the prong *a* moves from one extremity of its arc a' to the other a'', it sends a condensation down the tube; this condensation striking the surface of the water, is reflected by it up the tube. Now suppose that the front of this reflected condensation should just reach the prong at the instant it is starting on its retreat from a'' to a'; then the reflected condensation will conspire with the condensation formed by the prong in its retreat to make a greater condensation in the air outside the tube. Again, the retreat of the prong from a'' to a' produces in its rear a rarefaction, which also runs down the tube, is reflected, and will reach the prong at the instant it is about to return from a' to a'', and to cause a rarefaction in its rear; these two rarefactions moving in the same direction conspire to produce an intensified rarefaction. The original sound-waves thus combine with the reflected, to produce resonance; but this can only happen when the like parts of each wave coincide each with each; for if the tube were somewhat longer or shorter than it is, it is plain that condensations would meet rarefactions in the tube, and tend to destroy one another.

The loudness of sound of all wind instruments is due to the resonance of the air contained within them. A simple vibratory movement at the mouth or orifice of the instrument, scarcely audible in itself, such as the vibration of a reed in reed pipes, or a pulsatory movement of the air produced by the passage of a thin sheet of air over a sharp wooden or metallic edge, as in organ pipes, flutes, and flageolets, or more simply still by the friction of a gentle stream of breath from the lips sent obliquely across the open end of a closed tube, bottle, or pen-case, is sufficient to set the large body of enclosed air in the instrument into vibration, and thus reenforced, the sound becomes audible at long distances.

Experiment 206. — Attach a rose gas-burner A (Fig. 220) to a metal gas-tube about 1^m in length, and connect this by a rubber tube with a gas-burner. Light the gas at the rose burner, and you will hear a low, rustling noise. Remove the conical cap from the long tin tube (Fig. 215), support the tube in a vertical position, and gradually raise the burner into the tube; when it reaches a certain point not far up, the body of air in the tube will catch up the vibrations, and give out deafening sound-waves that will shake the walls and furniture in the room.

239. Measuring Wave-Lengths and the Velocity of Sound-waves. — Experiments like that described on page 253 enable us readily to measure the wave-length produced by a fork that makes a given number of vibrations in a second, and also to measure the velocity of sound-waves. It is evident that if a condensation generated by the prong of the fork in which its forward movement from a' to a'' (Fig. 220) met with no obstacle, its front, meantime, would traverse the distance od, or twice the distance oc; hence the length of the condensation is the distance od. But a condensation is only one-half of a wave, and the passage of the prong from a' to a'' is only one-half of a vibration; consequently the distance od is one-half of a wave-length, and the distance oc is one-fourth of a wave-length. The measured distance of oc in this case is about 13.13 inches; hence the length of wave produced by a C'-fork making

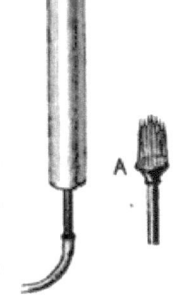

Fig. 220.

256 vibrations in a second is (13.13 inches × 4 =) 52.5 inches = 4.38 feet. And since a wave from this fork travels 4.38 feet in $\frac{1}{256}$ of a second, it will travel in an entire second (4.38 feet × 256 =) 1121 feet. The distance *oc* varies with the temperature of the air.

It is evident that the three quantities expressed in the formula

$$\text{wave-length} = \frac{\text{velocity}}{\text{number of vibrations}}$$

bear such a relation to one another that if any two are known, the remaining quantity can be computed. It will further be observed that *with a given velocity the wave-length varies inversely as the number of vibrations;* i.e. the greater the number of vibrations per second, the shorter the wave-length.

240. Interference of Sound-Waves.

Experiment 207. — Hold a vibrating diapason over a resonance-jar as in Figure 221. Roll the diapason over slowly in the fingers. At certain points, a quarter of a revolution apart, when the diapason is in an oblique position with reference to the edge of the jar as represented in the figure, the reënforcement from the tube almost entirely disappears, but reappears at the intermediate points. Return to the position where there is no resonance, and enclose in a loose roll of paper, the prong farthest from the tube, without touching the diapason, so as to prevent the sound-waves produced by that prong from passing into the tube; the resonance resulting from the vibrations of the other prong immediately appears.

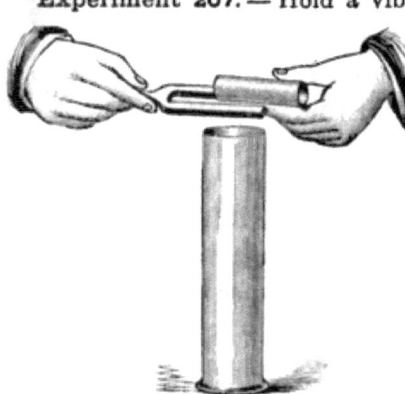

Fig. 221.

Experiment 208. — Select two of the tubes (Fig. 235) of nearly the same length, blow through them, and notice the peculiar throbbing sound produced by the interference of the two sounds.

Experiment 209. — Stop one of the orifices of a bicyclist's whistle (Fig. 222), and sound one whistle at a time. The sound of each is clear and smooth. Sound both whistles at the same time, and you obtain the usual rough and discordant sound.

Fig. 222.

The two whistles of unequal length give out waves of slightly different length, so that at certain short intervals the same phases of both sets will coincide (*i.e.* condensation with condensation) and produce intensified sounds which are heard at long distances, while at other intervals opposite phases coincide (*i.e.* condensation with rarefaction), and the result of their mutual destruction is to cause the otherwise smooth sound to become broken or rattling.

Two sound-waves may unite to produce a sound louder or weaker than either alone would produce, or even cause silence.

241. Forced and Sympathetic Vibrations.

Experiment 210. — Suspend from a frame several pendulums, A, B, C, etc. (Fig. 223). A and D are each 3 feet long, C a little longer, and B and E are shorter. Set A in vibration, and slight impulses will be communicated through the frame to D and cause it to vibrate. The vibration-period of D being the same as that of A, all the impulses tend to accumulate motion in D, so that it soon vibrates through arcs as large as those of A. On the other hand, C, B, and E, having different rates of vibration from that of A, will at first acquire a slight motion, but soon their vibrations will be in opposition to those of A, and then the impulses received from A will tend to destroy the slight motion they had previously acquired.

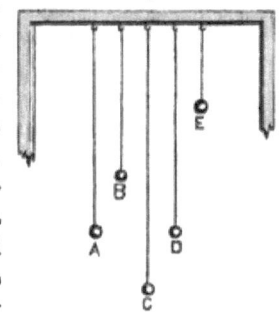

Fig. 223.

Experiment 211. — Press down gently one of the keys of a piano so as to raise the damper without making any sound, and then sing loudly into the instrument the corresponding note. The string corresponding to this note will be thrown into vibrations that can be heard for several seconds after the voice ceases. If another note be sung, this string will respond only feebly.

Raise the dampers from all the strings of the piano by pressing the foot on the right-hand pedal, and sing strongly some note into the piano. Although all the strings are free to vibrate, only those will respond loudly that correspond to the note you sing, *i.e.* those that are capable of making the same number of vibrations per second as are produced by your voice.

These experiments show that a vibrating body tends to make other bodies near it vibrate even if their periods of vibrations are different. Vibrations of this kind, such, for example, as those of B, C, and E in Experiment 210 and those generated in the sounding-boards of pianos, violins, etc., are called *forced vibrations*. But if the period of the incident waves of air is the same as that of the body which they cause to vibrate, the amplitude and intensity of the vibrations become very great, like that of the pendulum D, and those of the piano strings which gave forth the loud sounds. Such are called *sympathetic vibrations*.

QUESTIONS.

1. Why do not sound-waves travel with the same velocity through all bodies?

2. How are echoes produced?

3. On a day when sound-waves travel through the air at the rate of 1120 feet per second, what is the length of the sound-waves that proceed from a church bell which makes 192 vibrations in a second?

4. With what velocity do sound-waves travel when a jar whose depth is 10 inches gives the maximum reënforcement for a diapason which makes 256 vibrations in a second?

5. Great danger often arises from vibrations of the walls of a building caused by certain vibratory movements of machinery within. The danger in such cases can frequently be greatly diminished by changing the rate of motion in the machinery. Explain.

Section VII.

PITCH OF MUSICAL SOUNDS.

242. On What Pitch Depends.

Experiment 212. — Draw the finger-nail or a card slowly, and then rapidly, across the teeth of a comb. The two sounds produced are commonly described as *low* or *grave*, and *high* or *acute*. The hight of a musical sound is its *pitch*.

Experiment 213. — Cause the circular sheet-iron disk A (Fig. 224) to rotate, and hold a corner of a visiting-card so that at each hole an audible tap shall be made. Notice that when the separate taps or noises cease to be distinguishable, the sound becomes musical; also, that the pitch of the musical sound depends upon the rapidity of the rotation, *i.e.* upon the frequency of the taps.

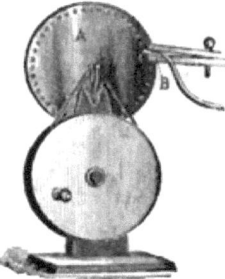

Fig. 224.

Experiment 214. — Hold the orifice of a glass tube B so as to blow through the holes as they pass. When rotating slowly, separate puffs are heard, from which it hardly seems possible to construct a musical sound. When, however, the ear is no longer able to detect the separate puffs, the sound becomes quite musical, and the pitch rises and falls with the speed.

Pitch depends upon frequency of vibration, or wave-length; i.e. *the greater the number of vibrations per second, or the shorter the wave-length, the higher the pitch.*

243. Musical Scale. — The pitch of a sound produced by twice as many vibrations as that of another sound is called the *octave* of the latter. Between two such sounds the voice rises or falls in a manner very pleasing to the ear by a definite number of steps. This gives rise to the

260   SOUND.

so-called *musical scale*, or *gamut*. The number of vibrations which shall constitute a given note is purely arbitrary, and differs slightly in different countries; but the ratios between the vibration numbers of the several notes of the gamut and the vibration number of the first or fundamental note of the gamut, are the same among all enlightened nations. The vibration numbers given in Figure 225 correspond to those of German instruments. For example, the string corresponding to the middle C (the key at the left of the two black keys near the middle of the key-board) of a German piano makes 264 vibrations in a second.

Notes.	Vibration numbers.	Vibration ratios.
C	132	1
D	148½	9/8
E	165	5/4
F	176	4/3
G	198	3/2
A	220	5/3
B	247½	15/8
C'	264	2
D'	297	9/4
E'	330	5/2
F'	352	8/3
G'	396	3
A'	440	10/3
B'	495	15/4
C''	528	4

Fig. 225.

Section VIII.

VIBRATION OF STRINGS.

244. Sonometer.

Experiment 215.—Stretch an elastic wire a over the bridges of the *sonometer* (Fig. 226), so that the portion between will be free

Fig. 226.

to vibrate. Pluck the string at its middle with the thumb and finger, causing it to vibrate, and observe the pitch. Next place a movable bridge d half-way between the two fixed bridges and cause the portion between either fixed bridge and the movable bridge to vibrate, and observe the change in pitch. How is the vibration period changed?

Experiment 216.— Stretch another wire *b*, either thicker or thinner than the last, employing the same length and tension as before, and notice the change in pitch due to the difference of weight of the wire. How is the vibration period changed?

Experiment 217.— Increase the tension of either wire by turning the pin, to which one end of the wire is attached, with a wrench C, and observe the change in pitch caused by change of tension. How does an increase of tension affect the vibration period?

Careful experiments show that *the vibration numbers of strings of the same material vary inversely as their lengths and the square roots of their weights, and directly as the square roots of their tension.*

245. Beats.

Experiment 218.— Strike simultaneously the lowest note of a piano and its sharp (black key next above), and listen to the resulting sound.

You hear a peculiar wavy or throbbing sound, caused by an alternate rising and sinking in loudness. These alternations in loudness are called *beats*.

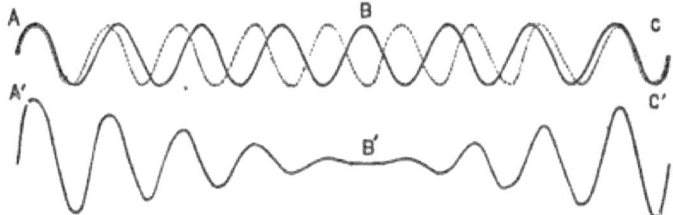

Fig. 227.

Let the continuous curve line AC (Fig. 227) represent a series of waves caused by striking the lower key, and the dotted line a series of waves proceeding from the upper key. Now the waves from both keys may start together at A; but as the waves from the lower key are given less frequently, so are they correspondingly longer; and at certain intervals, as at B, condensations will correspond with rarefactions, producing by their interference momentary silence, too short, however, to be perceived; but the sound as perceived by the ear is correctly represented in its varying loudness by the curved line in the lower part of the figure.

The number of beats per second due to two simple tones is equal to the difference of their respective vibration numbers. The sensation produced on the ear by such a throbbing sound, when the beats are sufficiently frequent, is unpleasant, much as the sensation produced by flashes of light that enter the eye, when you walk on the shady side of a picket fence, is unpleasant. The unpleasant sensation, called by musicians *discord*, is due to beats.

Section IX.

OVERTONES AND HARMONICS.

246. Vibration in Parts.

Experiment 219. — Hang up a rubber cord AC (Fig. 228) 4 feet long, and fasten both ends. Pluck it near the middle, and it will swing to and fro as a whole (2), at a rate dependent on its length, tension, etc. Hold it fast at B (3), and pluck it at a point half-way between A and B. Both halves are thrown into independent vibrations, and continue so to vibrate for a brief time after the hand is withdrawn from B. Again hold it fast at B, one-third its length above A (4), and pluck it half-way between A and B; the length BC instantly divides itself at B′ into two equal parts, and on withdrawing the hand from B, the whole cord is seen to vibrate in three distinct and equal sections. In a similar manner it may be made to vibrate in four, five, etc., sections.

OVERTONES AND HARMONICS.

Sounds coming from a string or other body that vibrates in parts are called *overtones*. If, as is the case with a string, the vibration number of the overtone is just two, three, four, etc., times that of the fundamental or lowest tone, the sound is called a *harmonic*. Many overtones can be produced from a steel bar or a metallic plate, but no harmonics. This distinction is of great importance, for, practically, no musical instruments are of much use unless their vibrating parts furnish harmonics.

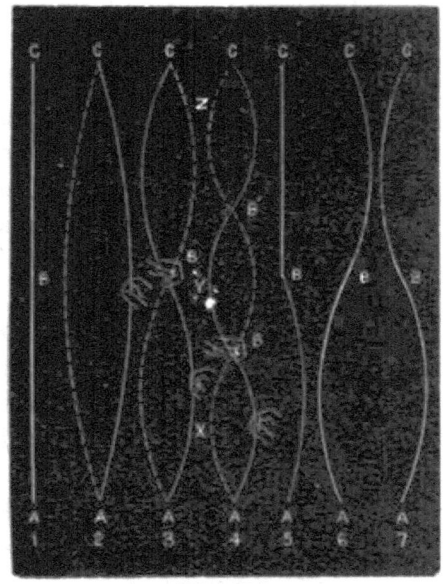

Fig. 228.

Experiment 220. — Press down the C′-key (middle C) of a piano gently, so that it will not sound; and while holding it down, strike the C-wire strongly. In a few seconds release the key, so that its damper will stop the vibrations of the string that was struck, and you will hear a sound which you will recognize by its pitch as coming from the C′-wire. Place your finger lightly on the C′-wire, and you will find that it is indeed vibrating. Press down the right pedal with the foot, so as to lift the dampers from all the wires, strike the C-key, and touch with the finger the C′-wire; it vibrates. Touch the keys next to C′, viz. B and D′; they have only a slight forced vibration. Touch G′; it vibrates.

Now it is evident that the vibrations of the C′ and G′-wires are sympathetic. A C-wire vibrating as a whole cannot cause sympathetic vibrations in a C′-wire; but if it vibrates in halves, it may. Hence we conclude that when the C-wire was struck, it vibrated, not only as a whole, giving a sound of its own pitch, but also in halves; and the result of this latter set of vibrations was, that an additional sound was produced by this wire, just an octave higher than the first-mentioned sound.

Again, the G'-wire makes three times as many vibrations as are made by the C-wire; hence the latter wire, in addition to its vibrations as a whole and in halves, must have vibrated in thirds, inasmuch as it caused the G'-wire to vibrate. It thus appears that a string may vibrate at the same time as a whole, in halves, thirds, etc., and the result is that *a sound is produced that is compounded of several sounds of different pitch.*

Not only do stringed instruments produce compound tones, but no ordinary musical instrument is capable of producing a *simple tone, i.e.* a sound generated by vibrations of a single period. In other words, *when any note of any musical instrument is sounded, there is produced, in addition to the primary tone, a number of other tones in a progressive series, each tone of the series being usually of less intensity than the preceding.* The primary or lowest tone of a note is usually sufficiently intense to be the most prominent, and hence is called *the fundamental tone.*

That two notes sounded together may harmonize, it is essential not only that the pitch of their fundamental tones be so widely different that they cannot produce audible beats, but that no beat shall be formed by their overtones, or by an overtone and a fundamental. Not only is there perfect agreement among the overtones of two notes an octave apart when sounded together, as when male and female voices unite in singing the same part of a melody, but the richness and vivacity of the sound is much increased thereby.

Section X.

QUALITY OF SOUND.

247. How Sounds from Different Sources are Distinguished. — We easily learn to distinguish by certain peculiarities the voices of our acquaintances. So we readily distinguish sounds emanating from various musical instruments, *e.g.* a piano, violin, harp, and cornet. It is not necessarily by the loudness or pitch of the sounds that we recognize them. It is by another property of sound called *quality*. *Two sounds can differ from each other in only three particulars*, viz. *intensity, pitch, and quality.*

Pitch depends on frequency of vibrations, loudness on their amplitude; *on what does quality depend?*

248. Analysis of Sounds. — The unaided ear is unable, except to a very limited extent, to distinguish the individual tones that compose a note. Helmholtz arranged a series of resonators consisting of hollow spheres of brass, each having two openings: one (A, Fig. 229) large, for the reception of the sound-waves,

Fig. 229.

and the other (B) small and funnel-shaped, and adapted for insertion into the ear. Each resonator of the series was adapted by its size to resound powerfully to only a single tone of a definite pitch. When any musical sound is produced in front of these resonators, the ear, placed at the orifice of any one, is able to single out from a collection that overtone, if present, to which alone this resonator is capable of responding. In this manner a complete analysis of any musical sound may be made, and the pitch and intensity of each of its components determined.

It is found that when a note is produced on a given instrument, not only is there a great variety of intensity represented by the overtones, but all the possible overtones of the series are by no means present. Which are wanting depends very much, in stringed instruments, upon the point of the string struck. For example, if a string is struck in its middle, no node can be formed at that point; consequently, the two important overtones produced by 2 and 4 times the number of vibrations of the fundamental will be wanting. Strings of pianos, violins, etc., are generally struck near one of their ends, and thus they are deprived of only some of their higher and feebler overtones.

249. Synthesis of Sounds. — The sound of a tuning-fork, when its fundamental is reënforced by a suitable resonance-cavity, is very nearly a simple tone. By sounding simultaneously several forks of different but appropriate pitch, and with the requisite relative intensities, Helmholtz succeeded in producing sounds peculiar to various musical instruments, and even in imitating most of the vowel sounds of the human voice.

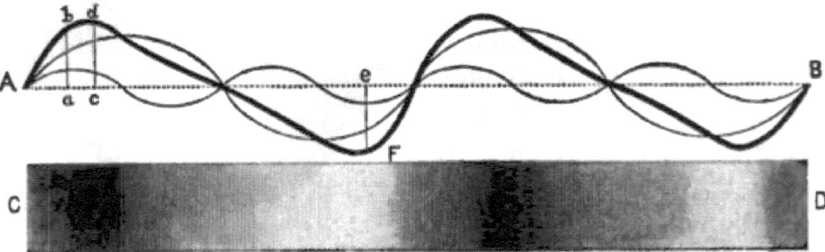

Fig. 230.

Thus it appears that he has been able to determine, both analytically and synthetically, that *the quality of a given sound depends upon what overtones combine with its fundamental tone, and on their relative intensities;* or, we may say more briefly, *upon the form of vibration*, since the form must be determined by the character of its components.

Section XI.

COMPOSITION OF SONOROUS VIBRATIONS, AND THE RESULTANT WAVE-FORMS.

250. Method of Representing Sound-Vibrations Graphically. — It is evident that there must be a particular aërial wave-form corresponding to each compound vibration, otherwise the ear would not be able to appreciate a difference in the quality of sounds to which these combination forms give rise. Every particle of air engaged in

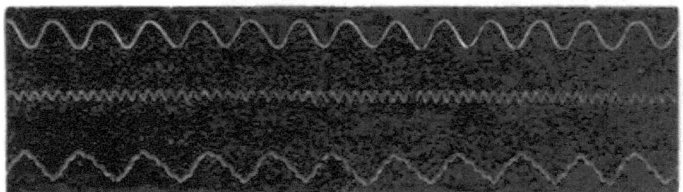

Fig. 231.

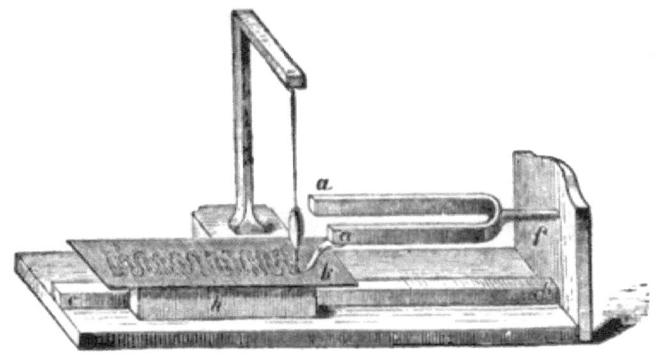

Fig. 232.

transmitting a compound sound-wave is simultaneously acted upon by several sets of vibratory movements, and it remains to investigate what its motion will be under their joint influence.

The light wave-lines AB (Fig. 230) represent typically two series of aerial sound-waves, corresponding respectively to a fundamental tone and its first overtone. The heavy line represents the form of the joint wave which results from the combination of the two constituents. If we suppose lines perpendicular to the axis, that is, to the dotted line, or line of repose, to be drawn to each point in this line, as *ab*, *cd*, *eF*, etc., they will represent by their varying lengths the displacement of any particle in a vibrating body, or any particle of air traversed by sound-waves, from its normal position.

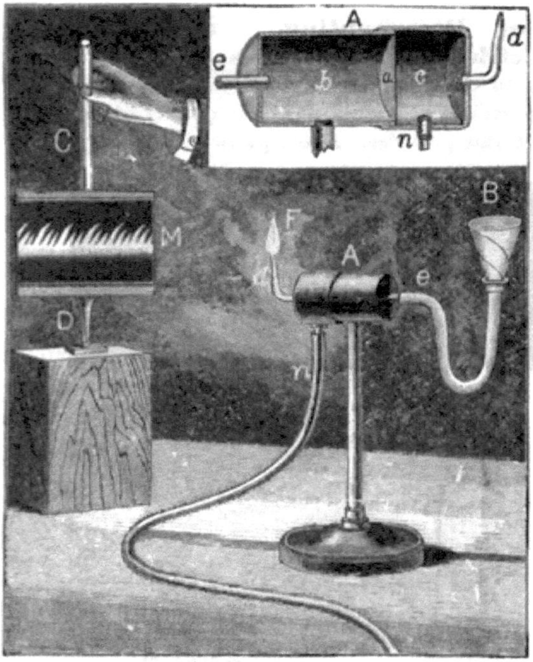

The rectangular diagram CD is intended to represent a portion of a transverse section of a body of air traversed by the joint wave represented by the heavy wave-line above. The depth of shading in different parts indicates the degree of condensation at those parts.

Figure 231 represents wave-lines drawn by an instrument called a *vibrograph* (Fig. 232). The second line represents a sound two octaves above that which the first line represents, and the third line shows the result of the combination of the two sets of vibrations.

Fig. 233.

251. Manometric Flames. — Apparatus like that shown in Figure 233 will serve to illustrate in a pleasing manner many facts pertaining to sound vibrations.

The cylindrical box A is divided by a membrane *a* into two compartments *c* and *b*. Illuminating-gas is introduced into the compartment *c*, through the rubber tube *n*, and burned at the orifice *d*. CD is a frame holding two mirrors, M, placed back to back, so that whichever side is turned toward the flame there is a reflection of the flame.

COMPOSITION OF SONOROUS VIBRATIONS. 269

When the mirror is at rest, an image of the flame will appear in the mirror as represented by A (Fig. 234). If the mirror is rotated, the flame appears drawn out in a band of light, as shown in B of the same figure.

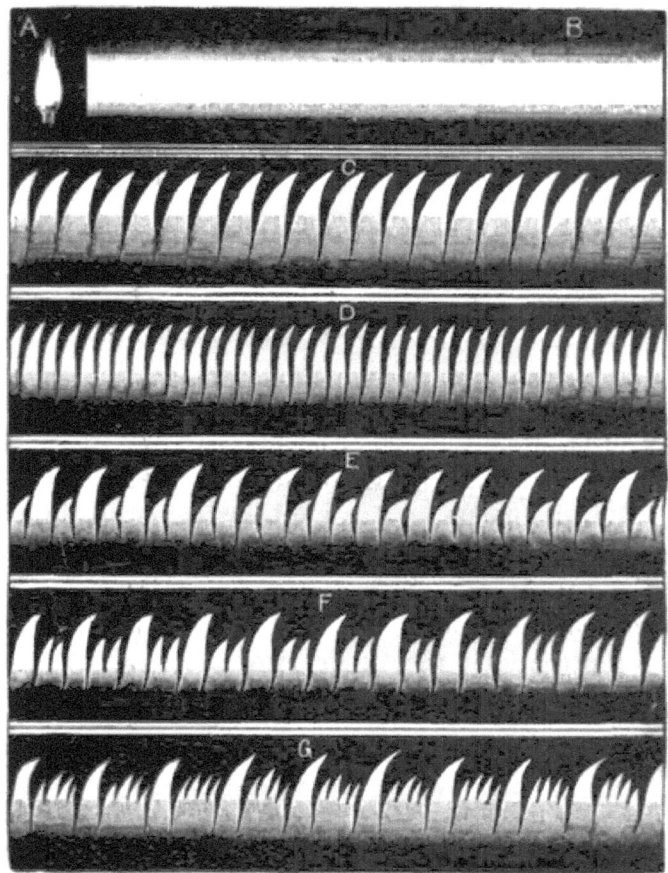

Fig. 234.

Sing into the cone B (Fig. 234) the sound of *oo* in tool, and waves of air will run down the tube, beat against the membrane *a*, causing it to vibrate, and the membrane in turn acts upon the gas in the compartment *c*, throwing it into vibration. The result is, that instead of a flame appearing in the rotating mirror as a continuous band of light, as B, Figure 234, it is divided up into a series of tongues of light, as shown in C, each condensation being represented by a tongue, and each rarefaction by a dark interval between the tongues. If a note an octave higher than the last is sung, we obtain, as we should expect, twice as many tongues in the same space, as shown in D. E represents the result when the two tones are produced simultaneously, and illustrates in a striking manner the effect of interference. F represents the result when the vowel *e* is sung on the key of C′; and G, when the vowel *o* is sung on the same key. These are called *manometric flames*.

Section XII.

MUSICAL INSTRUMENTS.

252. Classification of Musical Instruments. — Musical instruments may be grouped into three classes: (1) stringed instruments; (2) wind instruments, in which the sound is due to the vibration of columns of air confined in tubes; (3) instruments in which the vibrator is a membrane or plate. The first class has received its share of attention; the other two merit a little further consideration.

253. Wind Instruments.

Experiment 221. — Figure 235 represents a set of Quinke's whistles. The tubes are of the same size, but of varying length. Blow through the small tube across the lips of the large tube of each whistle in the order of their lengths, commencing with the longest.

Repeat the experiment, closing the end of the whistle farthest from you with a finger, so as to make what is called a "closed pipe."

The pitch of vibrating air-columns, as well as of strings, varies with the length, and *in both stopped and open pipes the number of vibrations is inversely proportional to the length of the pipe. An open pipe gives a note an octave higher than a closed pipe of the same length.*

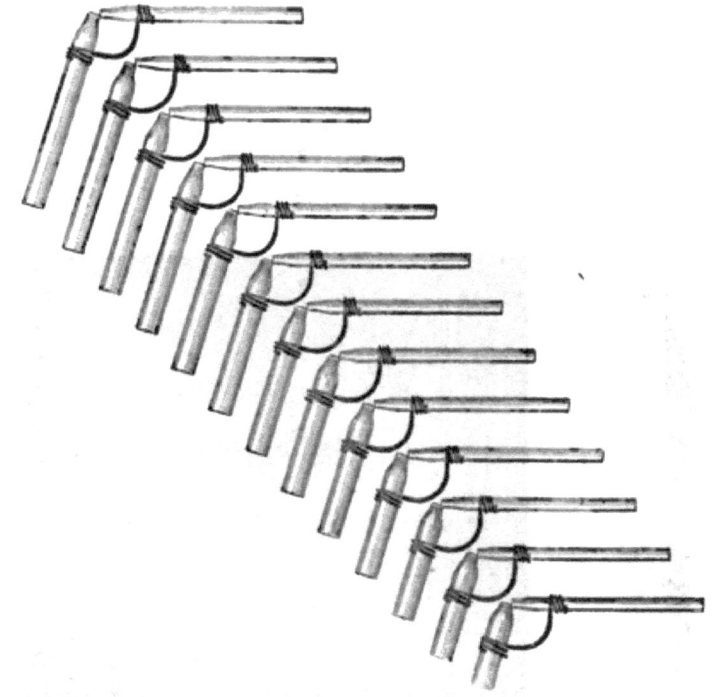

Fig. 235.

Experiment 222. — Take some of the longer whistles, blow as before, gradually increasing the force of the current. It will be found that only the gentle current will give the full musical fundamental tone of the tube, — a little stronger current produces a mere rustling sound; but when the force of the current reaches a certain limit, an overtone will break forth; and, on increasing still further the power of the current, a still higher overtone may be reached.

Figure 236 represents an open organ-pipe provided with a glass window A in one of its sides. A wire hoop B has stretched over it a membrane, and the whole is suspended by a thread within the pipe. If the membrane is placed near the upper end, a buzzing sound proceeds from the membrane when the fundamental tone of the pipe is sounded; and sand placed on the membrane will dance up and down in a lively manner. On lowering the membrane, the buzzing sound becomes fainter, till, at the middle of the tube, it ceases entirely, and the sand becomes quiet. Lowering the membrane still further, the sound and dancing recommence, and increase as the lower end is approached.

When the fundamental tone of an open pipe is produced, its air-column divides itself into two equal vibrating sections, with the anti-node at the extremities of the tube, and a node in the center.

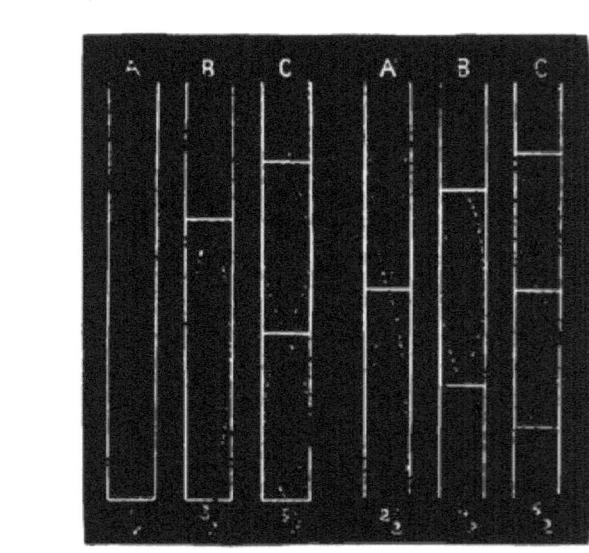

Fig. 236.   Fig. 237.

If the pipe is stopped, there is a node at the stopped end; if it is open, there is an anti-node at the open end; and in both cases there is an anti-node at the end where the wind enters, which is always to a certain extent open.

A, B, and C of Figure 237 show respectively the positions of the nodes and anti-nodes for the fundamental tone and first and second overtones of a closed pipe; and A', B', and C' show the positions of the same in an open pipe of the same length. The distance between the dotted lines shows the relative amplitudes of the vibrations of the air-particles at various points along the tube. Now the distance between a node and the nearest antinode is a quarter of a wave-length. Comparing, then, A and A', it will be seen that the wave-length of the fundamental of the closed pipe must be twice the wave-length of the fundamental of the open pipe; hence the vibration period of the latter is half that of the former; consequently the fundamental of the open pipe must be an octave higher than that of the closed pipe.

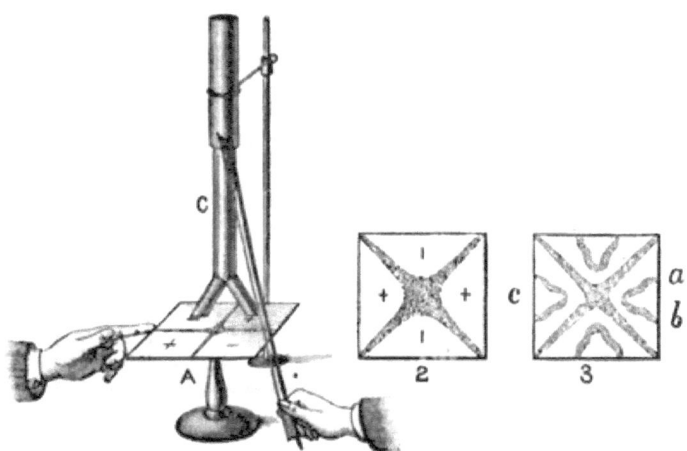

Fig. 238.

254. Sounding Plates, etc.

Experiment 223. — Fasten with a screw the elastic brass plate A (Fig. 238) on the upright support. Strew writing-sand over the plate, and draw a rosined bass bow steadily and firmly over one of its edges near a corner; and at the same time touch the middle of one

274 SOUND.

of its edges with the tip of the finger; a musical sound will be produced, and the sand will dance up and down, and quickly collect in two rows, extending across the plate at right angles to one another. Draw the bow across the middle of an edge, and touch with a finger one of its corners; the sand will arrange itself in two diagonal rows (2) across the plate, and the pitch of the note will be a fifth higher. Touch, with the nails of the thumb and forefinger, two points a and b (3) on one edge, and draw the bow across the middle c of the opposite edge, and you will obtain additional rows and a shriller note.

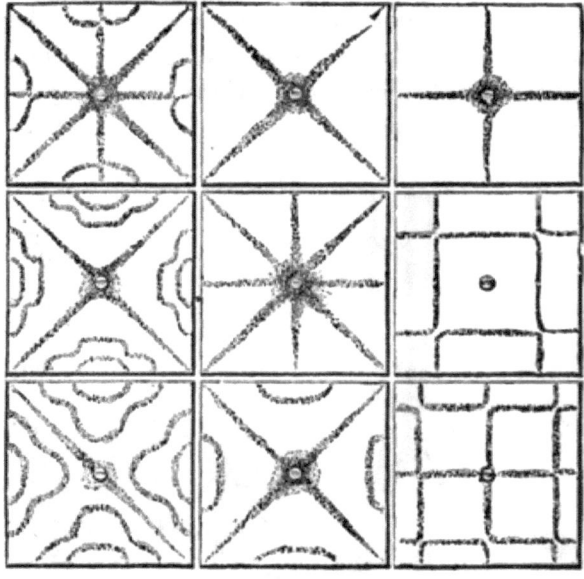

Fig. 239.

By varying the position of the point touched and bowed, a great variety of patterns can be obtained, some of which are represented in Figure 239. It will be seen that the effect of touching the plate with a finger is to prevent vibration at that point, and consequently a node is there produced. The whole plate then divides itself up into segments with nodal division lines in conformity with the node just formed. The sand rolls away from those parts which are alternately thrown into crests and troughs, to the parts that are at rest.

255. Interference.

Experiment 224. — C (Fig. 238) is a tin tube made in two parts to telescope one within the other. The extremity of one of the parts terminates in two slightly smaller branches. Bow the plate, as in the first experiment (1), place the two orifices of the branches over the segments marked with the + signs, and regulate the length of the tube so as to reënforce the note given by the plate, and set the plate in vibration. Now turn the tube around, so that one orifice may be over a + segment, and the other over a − segment; the sound due to resonance entirely ceases. It thus appears that the two segments marked + pass through the same phases together; likewise the phases of − segments correspond with one another; *i.e.* when one + segment is bent upward, the other is bent upward, and at the same time the two − segments are bent downward; for, when the two orifices of the tube are placed over two + segments or two − segments, two condensations followed by two rarefactions pass up these branches and unite at their junction to produce a loud sound; but when one of the orifices is over a + segment, and the other over a − segment, a condensation passes up one branch at the same time that a rarefaction passes up the other, and the two destroy one another when they come together; *i.e.* the two sound-waves combine to produce silence.

256. Bells. — A bell or goblet is subject to the same laws of vibration as a plate.

Fig. 240.

Experiment 225. — Nearly fill a large goblet with water, strew upon the surface lycopodium powder, and draw a rosined bow gently across the edge of the glass. The surface of the water will become rippled with wavelets (Fig. 240) radiating from four points 90° apart, corresponding to the centers of four ventral segments into which the goblet is divided, and the powder will collect in lines proceeding from the nodal points of the bell. By touching the proper points of a bell or glass with a finger-nail, it may be made to divide itself, like a plate, into 6, 8, 10, etc. (always an even number), vibrating parts.

Experiment 226. — Remove the brass plate (Fig. 239) from its support, and fasten the bell B (Fig. 241) on the support. Bow the

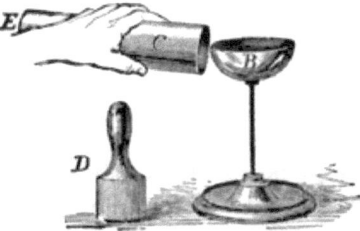

Fig. 241.

edge of the bell at some point, and hold the open tube C in a horizontal position with the center of one of its walls near that point of the edge of the bell which is opposite the point bowed. The tube loudly reënforces the sound of the bell. Move the tube around the edge of the bell and find its nodes.

Thrust the plunger D into the open end E of the tube, and find what part of the length of an open tube a closed tube should be to reënforce a sound of a given pitch.

257. Vocal Organs. — It is difficult to say which is more to be admired, — the wonderful capabilities of the

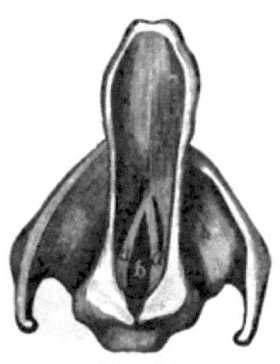

Fig. 242.

human voice or the extreme simplicity of the means by which it is produced. The organ of the voice is a reed instrument situated at the top of the windpipe, or trachea. A pair of elastic bands *aa* (Fig. 242), called the *vocal chords*, is stretched across the top of the windpipe. The air-passage *b*, between these chords, is open while a person is breathing; but when he speaks or sings, they are brought together so as to form a narrow, slit-like opening, thus making a sort of double reed, which vibrates when air is forced from the lungs through the narrow passage, somewhat like the little tongue of a toy trumpet. The sounds are grave or high according to the tension of the chords, which is regulated by muscular action. The cavities of the mouth and the nasal passages form a compound resonance-tube. This tube adapts itself, by its varying width and length, to the pitch of the note produced by the vocal chords. Place a finger on the protuberance of the throat called "Adam's apple," and sing a low note; then sing a high note, and you will observe that the protuberance rises in the latter case, thus shortening the distance between the vocal chords and the lips. Set a tuning-fork in vibration, open the mouth as if about to sing the corresponding note, place the fork in front of it, and the cavity of the mouth will resound to the note of the fork, but will cease to do so when the mouth adapts itself to the production of some other note. The different qualities of the different vowel sounds are produced by the varying forms of the resonating mouth-cavity, the pitch of the fundamental tones given by the vocal chords remaining the same. This constitutes *articulation*.

Section XIII.

SOME SOUND-WAVE RECEIVERS.

258. The Phonograph. — Figure 243 represents the Edison phonograph. A metallic cylinder A is rotated by means of a crank. On the surface of the cylinder is cut a shallow helical groove running around the cylinder from end to end, like the thread of a screw. A small metallic point, or style, projecting from the under side of a thin metallic disk D (Fig. 244), which closes one orifice of the mouth-piece B, stands directly over the thread. By a simple device the cylinder, when the crank is turned, is made to advance just rapidly enough to allow the groove to keep constantly under the style. The cylinder is covered with tinfoil. The cone F is usually applied to the mouth-piece to concentrate the sound-waves upon the disk D.

Now, when a person directs his voice toward the mouth-piece, the aërial waves cause the disk D to participate in every motion made by the particles of air as they beat against it, and the motion of the disk is communi-

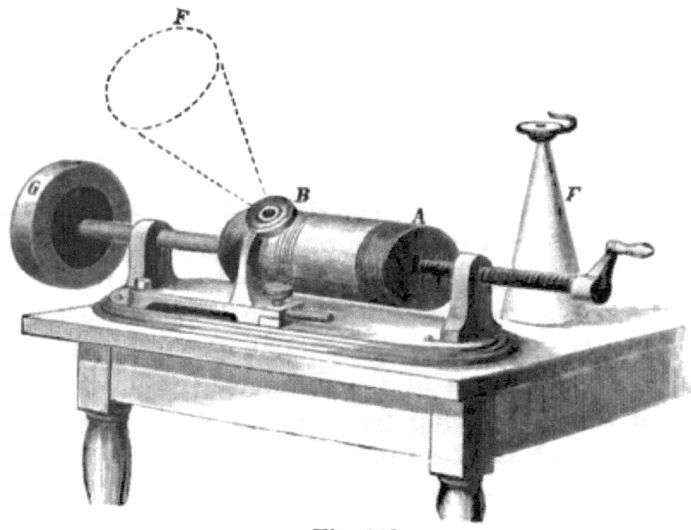

Fig. 243.

cated by the style to the tinfoil, producing thereon impressions or indentations as it passes on the rotating cylinder. The result is that there is left upon the foil an exact representation in relief of every movement made by the style. Some of the indentations are quite perceptible to the naked eye, while others are visible only with the aid of a microscope of high power. Figure 245 represents a piece of the foil as it would appear inverted after the indentations (here greatly exaggerated) have been imprinted upon it.

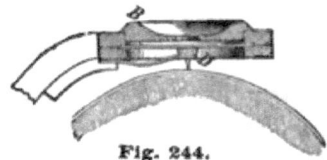

Fig. 244.

The words addressed to the phonograph having been thus impressed upon the foil, the mouth-piece and style are temporarily removed, while

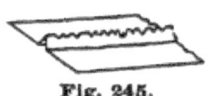

Fig. 245.

the cylinder is brought back to the position it had when the talking began, and then the mouth-piece is replaced. Now, evidently, if the crank is turned in the same direction as before, the style, resting upon the foil beneath, will be made to play up and down as it passes over ridges and sinks into depressions; this will cause the disk D to reproduce the same vibratory movements that caused the ridges and depressions in the foil. The vibrations of the disk are communicated to the air, and through the air to the ear; thus the words spoken to the apparatus may be, as it were, shaken out into the air again at any subsequent time, even centuries after, accompanied by the exact accents, intonations, and quality of sound of the original.

259. The Ear. — In Figure 246, A represents the external ear-passage; a is a membrane, called the *tympanum*, stretched across the bottom of the passage, and thus closing the orifice of a cavity b, called the *drum;* c is a

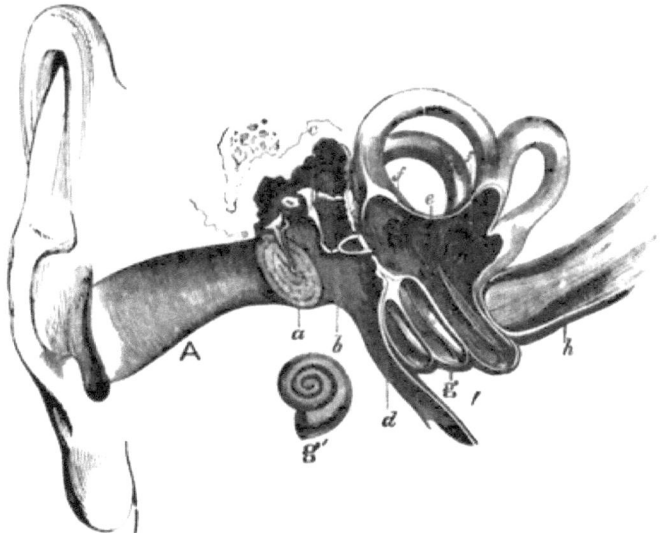

Fig. 246.

chain of small bones stretching across the drum, and connecting the tympanum with the thin membranous wall of the *vestibule* e; ff are a series of semicircular canals opening into the vestibule; g is the opening into another canal in the form of a snail-shell g', hence called the *cochlea* (this is drawn on a reduced scale); d is a tube (the *Eustachian tube*) connecting the drum with the throat; and h is the auditory nerve. The vestibule and all the canals opening into it are filled with a transparent liquid. The drum of the ear contains air, and the Eustachian tube forms a means of ingress and egress for air through the throat.

Now how does the ear hear? and how is it able to distinguish between the infinite variety of form, rapidity, and intensity of aërial sound-waves so as to interpret correctly the corresponding quality, pitch, and loudness of sound? Sound-waves enter the external ear-passage A as ocean-waves enter the bays of the seacoast, are reflected inward, and strike the tympanum. The air-particles, beating against this drum-head, impress upon it the precise wave-form that is transmitted to it through the air from the sounding body. The motion received by the drum-head is transmitted by the chain of bones to the membranous wall of the vestibule. From the walls of the spiral passage of the cochlea project into its liquid contents thousands of fine elastic threads or fibres, called "rods of Corti." As the passage becomes smaller and smaller, these vibratile rods become of gradually diminishing length and size (such as the wires of a piano may roughly represent), and are therefore suited to respond sympathetically to a great variety of vibration-periods. This arrangement is sometimes likened to a "harp of three thousand strings" (this being about the number of rods). The auditory nerve at this extremity is divided into a large number of filaments, like a cord unravelled at its end, and one of these filaments is attached to each rod. Now, as the sound-waves reach the membranous wall of the vestibule, they set it, and by means of it the liquid contents, into *forced vibration*, and so through the liquid all the fibres receive an impulse. Those rods whose vibration periods correspond with the periods of the constituents forming the compound wave are thrown into *sympathetic vibration*. The rods stir the nerve filaments, and the nerve transmits to the brain the impressions received. Just as a piano when its dampers are raised and a person sings into it, may be said to analyze each sound-wave, and show by the vibrating strings of how many tones it is composed, as well as their respective pitch, and by the amplitude of their vibrations their respective intensities; so, it is thought, this wonderful harp of the ear analyzes every complex sound-wave into a series of simple vibrations. Tidings of the disturbances are communicated to the brain, and there, in some mysterious manner, these disturbances are interpreted as sound of *definite quality, pitch, and intensity.*

CHAPTER VIII.

RADIANT ENERGY, ETHER-WAVES, — LIGHT.

Section I.

INTRODUCTION.

260. Energy Received from the Sun. — Exposed to the sun, the skin is warmed, — the sense of touch is affected; it is illuminated, — thereby the sense of sight is affected; it is tanned, — its chemical condition is changed. It is evident that we receive something which must come to us from the sun. To the sense of touch it appears to be heat; in the eye it produces the sensation of light; in certain substances it has the power to produce chemical changes. *What is it that we receive from the sun?*

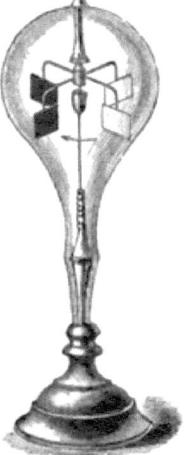

Fig. 247.

Figure 247 represents an instrument called a *radiometer*. The moving part is a small vane resting on the point of a needle. It is so nicely poised on this pivot that it rotates with the greatest freedom. To the extremities of each of the four arms of the vane are attached disks of aluminum, which are white on one side and black on the other. The whole is enclosed in a glass bulb, and the air within is reduced to less than one-millionth its usual density. If the instrument is exposed to the sun the wheel will rotate with the white faces in advance.

In just what manner it is caused to rotate does not concern us at present; but the fact that it rotates, and that it is caused to rotate directly or indirectly by something that comes from the sun, is pertinent to the question before us. Whenever a body is caused to move or increase its rate of motion, energy must be imparted to it; hence *energy must be imparted to the radiometer-vane by the sun.*

That which we receive from the sun, whether it affects the sense of touch or of sight, or produces chemical changes, is in reality some form of energy and is one and the same form whatever the effect.

261. Ether the Medium of Motion. — If we receive the energy of motion, what moves? Our atmosphere is but a thin mantle covering the earth, while the great space that separates us from the sun contains no air or other known substance. But *empty space cannot communicate motion*. It is assumed — *it is necessary to assume* — that there is some medium filling the interplanetary space; in fact, filling all space otherwise unoccupied, a medium by which motion can be communicated from one point to another. This medium has received the name of *ether*.

We cannot see, hear, feel, taste, smell, weigh, nor measure it. What evidence, then, have we that it exists? This: phenomena occur just as they *would* occur *if* all space were filled with an ethereal medium capable of transmitting motion; we have been able to account for these phenomena on no other hypothesis, hence our belief in the existence of the medium.

The transmission of energy through the medium of ether is called *radiation;* energy so transmitted is called *radiant energy*, and the body emitting energy in this manner is called a *radiator*.

262. Undulatory Theory; the Sensation of Light. — All evidence points to one conclusion: that we receive energy from the sun in the form of *vibrations* or *waves;* that a portion of these waves having suitable wave-length are capable of causing through the eye the sensation of *light*. Such as affect the sense of sight are called *light-waves*. This is known as the *undulatory theory*. According to this theory *light*[1] *is a sensation caused, usually, by the action of ether-waves on the organ of sight*. The term light is commonly applied to the agent which produces the sensation, but it is thought that in a scientific treatise much may be gained in many ways by restricting the term to the sensation, and applying to the agent the appropriate term *light-waves*.

All ether-waves are capable of generating heat and, consequently, of causing the sensation of *warmth*. A large portion of the ether-waves are also capable of promoting chemical action in certain substances.

263. Sources of Light-waves, Incandescence and Phosphorescence. — Every form of matter when sufficiently heated emits light-waves; in other words, when the vibration period of its molecules becomes such as to create ethereal waves that are capable of affecting the sense of sight, the body is said to be *luminous*. This condition is termed *incandescence*. The sun and fixed stars are in a condition of intense incandescence. Nearly all the artificial sources of light-waves, such as lamp and gas flames and electric lamps, depend upon the development of light-waves mainly through the incandescence of carbon.

---

[1] " The optical sensations are Light, Color, and Lustre." — Bain's *Mental Science*.

284 RADIANT ENERGY.

There is a class of substances, such as the sulphides of calcium, strontium, etc., which, after several hours' exposure to light-waves, absorb their energy (*i.e.* their molecules acquire sympathetic vibrations) without becoming hot, and in turn emit light-waves, which are quite perceptible in a dark room for several hours after the exposure. This property of shining in the dark after having been exposed to light-waves is termed *phosphorescence*. A so-called *luminous paint* is prepared and applied to certain parts of bodies that are exposed to sunshine during the day; at night those parts to which the paint is applied are alone luminous. This paint may be used for a variety of purposes, such as rendering danger signals, door numbers, and plates luminous (Fig. 248), etc.

Fig. 248.

264. Light-waves travel in Straight Lines. — The path of light-waves admitted into a darkened room through a small aperture, as indicated by the illuminated dust, is perfectly straight. *An object is seen by means of light-waves which it sends to the eye.* A small object placed in a straight line between the eye and a luminous point may intercept the light-waves in that path, and the point become invisible. Hence we cannot see around a corner, or through a bent tube.

265. Ray, Beam, Pencil. — Any line RR, Figure 249, which pierces the surface of an ether-wave *ab* perpendicularly is called a *ray*. The term "*ray*" *is but an expression for the direction in which motion is propagated, and along which the successive effects of ether-waves occur.*

If the wave-surface $a'b'$ is a plane, the rays $R'R'$ are parallel, and a collection of such rays is called a *beam*. If the wave-surface $a''b''$ is spherical or concave, the rays $R''R''$ have a common point at the center of curvature; and a collection of such rays is called a *pencil*.

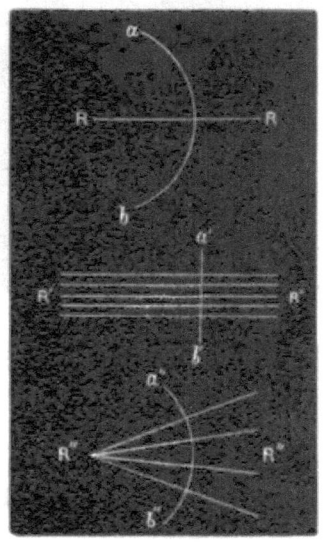

Fig. 249.

266. Transparent, Translucent, and Opaque Bodies. — Bodies are *transparent*, *translucent*, or *opaque*, according to the manner in which they act upon the light-waves which pass through them. Generally speaking, those objects are *transparent* that allow other objects to be seen through them distinctly, *e.g.* air, glass, and water. Those objects are *translucent* that allow light-waves to pass, but in such a scattered condition that objects are not seen distinctly through them, *e.g.* fog, ground glass, and oiled paper. Those objects are *opaque* that apparently cut off all the light-waves and prevent objects from being seen through them.

267. Luminous and Illuminated Objects. — Some bodies are seen by means of light-waves which they emit, *e.g.* the sun, a candle flame, and a "live" coal; they are called *luminous bodies*. Other bodies are seen only by means of light-waves which they receive from luminous ones; and when thus rendered visible are said to be illuminated, e.g. the moon, a man, a cloud, and a "dead" coal.

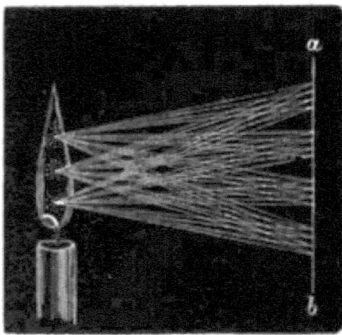

Fig. 250.

Every point of a luminous body is an independent source of light-waves, and emits light-waves in every direction. Such a point is called a *luminous point*. In Figure 250 there are represented a few of the infinite number of pencils emitted by three luminous points of a candle flame. Every point of an illuminated object, *ab*, receives light-waves from every luminous point.

268. Images formed through Small Apertures.

Experiment 227. — Cut a hole about 4 inches square in one side of a box; cover the hole with tin-foil, and prick a hole in the foil with a pin. Place the box in a darkened room, and a candle flame in the box near to the pin-hole. Hold an oiled-paper screen before the hole in the foil; an inverted image of the candle flame will appear upon the translucent paper.

An *image* is a kind of picture of an object. If light-waves from objects illuminated by the sun, *e.g.* trees, houses, clouds, or even an entire landscape, are allowed to pass through a small aperture in a window shutter and strike a white wall in a dark room, inverted images of the objects in their true colors will appear upon it. The cause of these phenomena is easily understood. When no screen intervenes between the candle and the screen A, Figure 251, every point of the screen receives light-waves from every point of the candle; consequently, on every point on A, images of the infinite number of points of the candle are formed. The result of the confusion of images is equivalent to no image. But let the screen B, containing a small hole, be interposed; then, since light-waves travel only in straight lines, the point Y' can only receive an image of the point Y, the point Z' only of the point Z, and so for intermediate points; hence a distinct image of the object must be formed on the screen A.

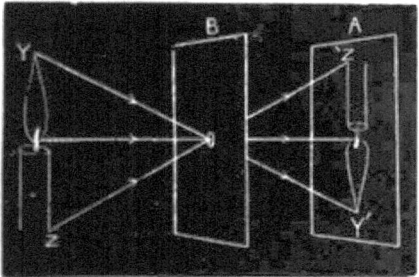

Fig. 251.

That an image may be distinct, the rays from different points of the object must not mix on the image, but all rays from each point on the object must be carried to its own point on the image.

269. Shadows.

Experiment 228. — Procure two pieces of tin or cardboard, one 18cm square, the other 3cm square. Place the first between a white wall and a candle flame in a darkened room. The opaque tin intercepts the light-waves that strike it, and thereby excludes light-waves from a space behind it.

This space is called a *shadow*. That portion of the surface of the wall that is darkened is a *section of the shadow*, and represents the form of a section of the body that intercepts the light-waves. A section of a shadow is frequently for convenience called a shadow. Notice that the shadow is made up of two distinct parts, — a dark center bordered on all sides by a much lighter fringe. The dark center is called the *umbra*, and the lighter envelope is called the *penumbra*.

Experiment 229.— Carry the tin nearer the wall, and notice that the penumbra gradually disappears and the outline of the umbra becomes more distinct. Employ two candle flames, a little distance apart, and notice that two shadows are produced. Move the tin toward the wall, and the two shadows approach one another, then touch, and finally overlap. Notice that where they overlap the shadow is deepest. This part gets no light-waves from either flame, and is a section of the umbra; while the remaining portion gets light-waves from one or the other, and is a section of the penumbra. Or move the eye across the shadow from side to side and see parts of the flame in the penumbra, but none in the umbra.

Just so *the umbra of every shadow is the part that gets no light-waves from a luminous body, while the penumbra is the part that gets light-waves from some portion of the body, but not from the whole.*

Experiment 230.— Repeat the above experiments, employing the smaller piece of tin, and note all differences in phenomena that occur. Hold a hair in the path of the sun's waves, about a quarter of an inch in front of a fly-leaf of this book, and observe the shadow cast by the hair. Then gradually increase the distance between the hair and the leaf, and note the change of phenomena.

If the source of light-waves were a single luminous point, as A (Fig. 252), the shadow of an opaque body B would be of infinite length, and would consist only of an umbra. But if the source of light-waves has a sensible size, the opaque body will intercept just as many separate pencils as there are luminous points, and consequently will cast an equal number of independent shadows.

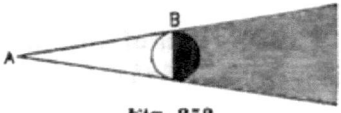
Fig. 252.

Let AB (Fig. 253) represent a luminous body, and CD an opaque body. The pencil from the luminous point A will be intercepted between the lines CF and DG, and the pencil from B will be intercepted between the wave-lines CE and DF. Hence the light-waves will be wholly excluded only from the space between the lines CF and DF, which enclose the umbra.

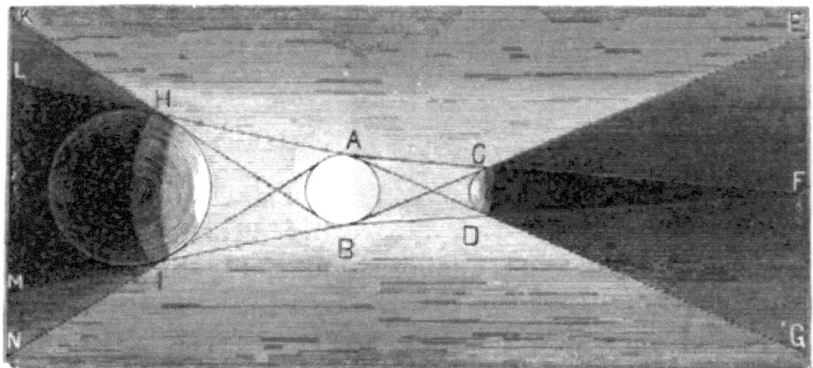

Fig. 253.

The enveloping penumbra, a section of which is included between the lines CE and CF, and between DF and DG, receives light-waves from certain points of the luminous body, but not from all.

Section II.

PHOTOMETRY, VISUAL ANGLE, ETC.

270. Law of Inverse Squares.

Experiment 231. — Arrange apparatus as follows: Draw a straight chalk-line across a table, and place at right angles to this line a row of four lighted candles, and on the same line, at a distance, a single lighted candle. Half-way between this candle and the row of candles place a paper disk having a circular translucent spot in the center, as in Figure 254. It is evident that one side of the paper receives four times the radiant energy that the other does. Move the row of candles slowly away from the paper, or move the single candle toward the paper, until a point is found where the spot nearly disappears. The paper now receives the same amount of energy from the single flame as from the four flames, but it will be found that the row of flames is twice as far from the paper as the single flame.

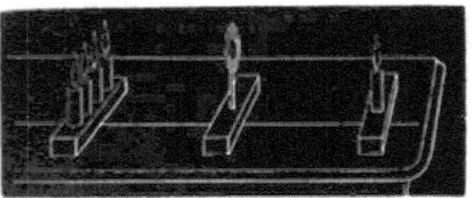

Fig. 254.

Thus, by doubling the distance, the intensity of illumination is diminished fourfold. In a similar manner it may be shown that at three times the distance it takes nine flames to be equivalent to one flame. Hence, *the intensity of illumination diminishes as the square of the distance increases.* This is called the *law of inverse squares.*

Experiment 232. — Introduce the paper disk, as above, between a candle flame and a kerosene or a gas flame, and so regulate the distance that the central spot will disappear; then calculate the relative intensities of the flames in accordance with the law of inverse squares.

This is the method usually employed by gas inspectors for testing the intensity of light-waves. Apparatus arranged for this purpose is called a *photometer.* "The *candle power*, which is the unit of intensity generally employed in photometry, is the intensity of the flame of a sperm candle weighing one-sixth of a pound, and burning one hundred and twenty grains an hour."

The relative brightness of the common sources of light-waves are approximately as follows: [1] —

Sun at its surface	190,000 candle power.
Most powerful electric arc	55,900 " "
Incandescent calcium	1,300 " "
Ordinary gas-burner	12 to 16 " "
Standard candle	1 " "

[1] C. A. Young.

271. Visual Angle.

Experiment 233. — Prick a pin-hole in a card, place an eye near the hole, and look at a pin about 20^{cm} distant. Then bring the pin slowly toward the eye, and the dimensions of the pin will appear to increase as the distance diminishes.

Why is this? We see an object by means of its image formed on the retina of the eye; and its apparent magnitude is determined by the extent of the retina covered by its image. Rays proceeding from opposite extremities of an object, as AB (Fig. 255), meet and cross one another

Fig. 255.

in the window of the eye, called the *pupil*. Now, as the distance between the points of the blades of a pair of scissors depends upon the angle that the handles form with one another, so the size of the image formed on the retina depends upon the size of the angle, called the *visual angle*, formed by these rays as they enter the eye. But the size of the visual angle diminishes as the distance of the object from the eye increases, as shown in the diagram; *e.g.* at twice the distance the angle is one-half as great; at three times the distance the angle is one-third as great; and so on. Hence, *distance affects the apparent size of an object.* Our judgment of size is, however, influenced by other things besides the visual angle which they subtend.

272. Velocity of Light-Waves. — By several ingenious methods it has been ascertained that light-waves travel at the rate of about 186,000 miles in a second, a velocity which would enable them to go around the earth about seven times in a second. Sound-waves travel in air at the rate of only about one-fifth of a mile per second. This great difference can be accounted for only on the supposition that *the rarity and elasticity of ether are enormously greater than that of air.*

Section III.

REFLECTION OF LIGHT-WAVES.

273. Law of Reflection.

Experiment 234. — Look through the hole in the metal band (Fig. 256), marked zero, at the mirror. You see in the mirror an image of the hole through which you are looking, but you do not see the image of any of the other holes. Rays that pass through this hole strike the mirror perpendicularly, and are called *incident* rays.

Fig. 256.

The *reflected* rays are thrown back in the same line and through the same hole that the incident rays travel to the eye.

Hold a candle flame at one of the other holes (or stop it with a finger), *e.g.* at the hole marked 10. You can see the reflected rays of the candle flame only through the hole of the same number on the other side, *i.e.* for example, incident rays making an angle of 10° (called the *angle of incidence*) with the perpendicular to the surface of the mirror is reflected at an angle of 10° (called the *angle of reflection*) with the perpendicular. *The angle of reflection is always equal to the angle of incidence.*

274. Reflection from Plane Mirrors; Virtual Images. — MM (Fig. 257) represents a plane mirror, and AB a pencil of divergent rays proceeding from the point A of an object AH. Erecting perpendiculars at the points of incidence, or the points where these rays strike the mirror, and making the angles of reflection equal to the angles of incidence, the paths BC and EC of the reflected rays are found.

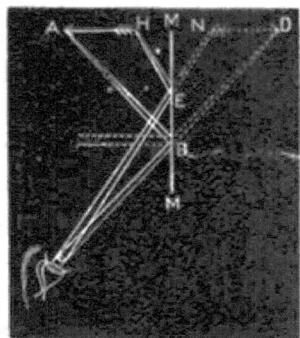

Fig. 257.

It appears that *divergent incident rays remain divergent after reflection from a plane mirror.* (In like manner construct a diagram, and show that *parallel incident rays are parallel after reflection.*) Construct another diagram, and show that *convergent incident rays are convergent after reflection, i.e.* reflection from a plane surface does not alter the angle between rays. To an eye placed at C, the points from which the rays appear to come are of course in the direction of the rays as they enter the eye. These points may be found by *continuing* the rays CB and CE behind the mirror, till they meet at the points D and N. Every point of the object AH sends out its pencil of rays; and those that strike the mirror at a suitable angle to be reflected to the eye, produce on the retina of the eye an image of that point, and the point from which the light-waves appear to emanate is found, as previously described. Thus, the pencils EC and BC appear to emanate from the points N and D; and the whole body of light-waves received by the eye seems to come from an *apparent object* ND behind the mirror. This apparent object is called an *image;* but as, of course, there can be no real image formed there, it is called a *virtual* or an *imaginary* image. It will be seen, by construction, that *an image in a plane mirror appears as far behind the mirror as the object is in front of it, and is of the same size and shape as the object.*

275. Reflection from Concave Mirrors. — Let MM' (Fig. 258), represent a section of a concave mirror, which may be regarded as a small part of a hollow spherical shell having a polished interior surface. The distance MM' is called the *diameter of the mirror.* C is the center

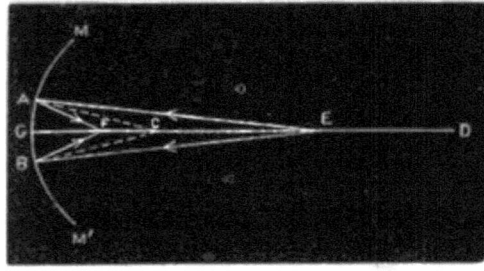

Fig. 258.

of the sphere, and is called the *center of curvature.* G is the *vertex* of the mirror. A straight line DG drawn through the center of curvature and the vertex is called the *principal axis* of the mirror. A concave mirror may be considered as made up of an infinite number of small plane surfaces. All radii of the mirror, as CA, CG, and CB, are perpendicular to the small planes which they strike. If C be a luminous point, it is evident that all light-waves emanating from this point, and striking the mirror, will be reflected to its source at C.

Let E be any luminous point in front of a concave mirror. To find the direction that rays emanating from this point take after reflection, draw any two lines from this point, as EA and EB, representing two of the infinite number of rays composing the divergent pencil that strikes the mirror. Next, draw radii to the points of incidence A and B, and draw the lines AF and BF, making the angles of reflection equal to the angles of incidence. Place arrow-heads on the lines representing rays to indicate the direction of the motion. The lines AF and BF represent the direction of the rays after reflection.

It will be seen that the rays after reflection are convergent, and meet at the point F, called the *focus*. This point is the focus of all reflected rays that emanate from the point E. It is obvious that if F were the luminous point, the lines AE and BE would represent the reflected rays, and E would be the focus of these rays. Since the relation between the two points is such that light-waves emanating from either one are brought by reflection to a focus at the other, these points are called *conjugate foci*. *Conjugate foci are two points so related that the image of either is formed at the other.* The rays EA and EB emanating from E are less divergent than rays FA and FB, emanating from a point F less distant from the mirror, and striking the same points. Rays emanating from D, and striking the same points A and B, will be still less divergent; and if the point D were removed to a distance of many miles, the rays incident at these points would be very nearly parallel. Hence rays may be regarded as practically parallel when their source is at a very great distance, *e.g.* the sun's rays. If a sunbeam, consisting of a bundle of parallel rays, as EA, DG, and HB (Fig. 259), strike a concave mirror parallel with its principal axis, these rays become convergent by reflection, and meet at a point (F) in the principal axis. This point, called the *principal focus*, is *just half-way between the center of curvature and the vertex of the mirror.*

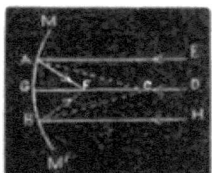

Fig. 259.

On the other hand, it is obvious that *divergent rays*

emanating from the principal focus of a concave mirror become parallel by reflection.

If a small piece of paper is placed at the principal focus of a concave mirror, and the mirror is exposed to the parallel rays of the sun, the paper will quickly burn.

Construct a diagram, and show that *rays proceeding from a point between the principal focus and the mirror are divergent after reflection, but less divergent than the incident rays.* Reversing the direction of the rays the same diagram will show that *convergent rays are rendered more convergent by reflection from concave mirrors.*

The general effect of a concave mirror is to increase the convergence or to decrease the divergence of incident rays.

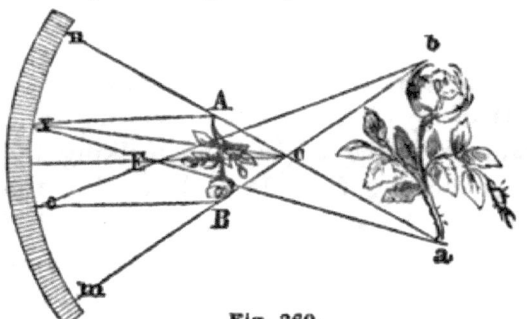

Fig. 260.

The statement, that parallel rays after reflection from a concave mirror meet at the principal focus, is only approximately true. The smaller the diameter of the mirror, the more nearly true is the statement. It is strictly true only of parabolic mirrors. Such are used in the head-lights of locomotives.

276. Formation of Images.

Experiment 235. — Hold some object, *e.g.* a rose, as *ab* (Fig. 260), a few feet in front of a concave mirror. Looking in the direction of the axis of the mirror you see a small inverted image AB of the object between the center of curvature, C, of the mirror and its principal focus F.

Evidently if AB represent an object placed between the principal focus and center of curvature, then *ab* will represent the *image* of the object. The image in this case may be projected upon a screen, but it will not be so bright as in the former case, because the light-waves are spread over a larger surface.

Experiment 236. — Place a candle in an otherwise dark room 20 feet from the mirror, catch the focused light-waves upon a paper screen, and show that the focus is half-way between the vertex and the center of curvature of the mirror.

Experiment 237. — Advance the distant candle flame toward the mirror, moving it up and down. (1) Show that the focus advances to meet the flame, and that when the flame is raised, the focus is depressed, and the converse. (2) Show that when the flame is at the center of curvature, there also is the focus. (3) Show that when the flame is between the center of curvature and the principal focus, the focus of the flame is farther away than the center of curvature. (4) Show that when the flame is at the principal focus, the reflected rays are parallel, or the focus is at an infinite distance. (5) Show that when the flame is still nearer, the reflected rays diverge and appear to come from a point behind the mirror. (6) Notice that in all cases except the last the images are real and inverted, and that in all cases where a real image is formed, the flame and the image may change places.

Experiment 238. — Form a real image of the flame between yourself and the mirror; view the image through a convex lens (Fig. 280); show that the image can be magnified by a convex lens, and thereby illustrate the principle of an astronomical reflecting telescope.

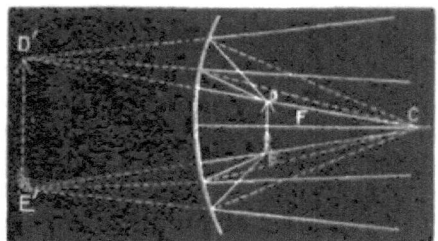

Fig. 261.

Construct the image of an object placed between the principal focus and the mirror, as in Figure 261. It will be seen in this case that a pencil of rays proceeding from any point of an object, *e.g.* D, has no actual focus, but appears to proceed from a *virtual* focus D′, back of the mirror; and so with other points, as E. *The image of an object placed between the principal focus and the mirror is virtual, erect, larger than the object, and is back of the mirror.*

277. Convex Mirrors. — *The general effect of convex mirrors is to separate incident rays.* In them all images are *virtual, erect, and smaller than the objects.*

Section IV.

REFRACTION.

278. Introductory Experiments.

Experiment 239. — Into a darkened room admit a sunbeam so that its rays may fall obliquely on the bottom of the basin (Fig. 262),

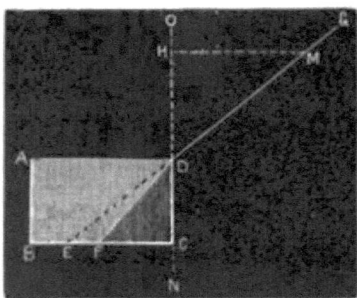

Fig. 262.

and note the place on the bottom where the edge of the shadow DE cast by the side of the basin DC meets the bottom at E. Then, without moving the basin, fill it even full with water slightly clouded with milk or with a few drops of a solution of mastic in alcohol. It will be found that the edge of the shadow has moved from DE to DF, and meets the bottom at F. Beat a blackboard rubber, and create a cloud of dust in the path of the beam in the air, and you will discover that the rays GD that graze the edge of the basin at D become bent at the point where they enter the water, and now move in the bent line GDF, instead of, as formerly, in the straight line GF. The path of the line in the water is now nearer to the vertical side DC; in other words, this part of the beam *is more nearly vertical than before.*

Experiment 240. — Place a coin (A, Fig. 263) on the bottom of an empty basin, so that, as you look through a small hole in a card BC over the edge of the vessel, the coin is just out of sight. Then, without moving the card or basin, fill the latter with water. Now, on looking through the aperture in the card, the coin is visible. The beam AE, which formerly moved in the straight line AD, is now bent at E, where it leaves the water, and, passing through the aperture in the card, enters the eye. Observe that, as the beam

Fig. 263.

passes from the water into the air, it is turned farther from a vertical line EF; in other words, *the beam is farther from the vertical than before.*

Experiment 241. — From the same position as in the last experiment, direct the eye to the point G in the basin filled with water. Reach your hand around the basin, and place your finger where that point appears to be. On examination, it will be found that your finger is considerably above the bottom. Hence, *the effect of the bending of rays, as they pass obliquely out of water, is to cause the bottom to appear more elevated than it really is;* in other words, *to cause the water to appear shallower than it is.*

Experiment 242. — Thrust a pencil obliquely into water; it will appear shortened, bent at the surface of the water, and the immersed portion elevated.

Experiment 243. — Place a piece of wire (Fig. 264) vertically in front of the eye, and hold a narrow strip of thick plate glass horizontally across the wire, so that the light-waves from the wire may pass obliquely through the glass to the eye. The wire will appear to be broken at the two edges of the glass, and the intervening section will appear to the right or left according to the inclination of the glass; but if the glass is not inclined to the one side or the other, the wire does not appear broken.

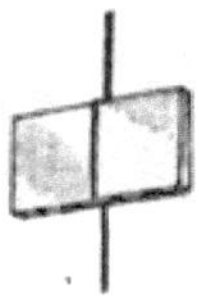

Fig. 264.

Experiment 244. — Partly fill the cell (Fig. 147) with carbon bisulphide, then add water. Place the cell in the path of a beam reflected from a *porte lumière*. Place vertically in front of the cell a wire, and project with a lens a shadow of the wire on a screen. Turn the cell obliquely, as in the last experiment, and notice the difference in the refracting power of the two liquids.

Experiment 245. — Partly fill the same cell with water. Focus it on the screen so that the surface of the water will be visible. Add a lump of ice on the water. Observe the streakiness caused by difference in the density of water at different temperatures.

Experiment 246. — Project with a lens a luminous circle on a screen. Hold, a few feet in front of the screen, a candle flame in the path of the light-waves. Observe the wavy streakiness arising from the changing density of the air and convection currents.

When a light-beam passes from one medium into another of different density, it is bent or *refracted* at the boundary plane between the two media, unless it falls exactly perpendicularly on this plane. *If it pass into a denser medium, it is refracted toward a perpendicular to this plane; if into a rarer medium, it is refracted from the perpendicular.* The angle GDO (Fig. 262) is called the *angle of incidence;* FDN, the *angle of refraction;* and EDF, the *angle of deviation.*

279. Cause of Refraction. — *Careful experiments have proved that the velocity of light-waves is less in a dense than in a rare medium.* Let the series of parallel lines AB (Fig. 265) represent a series of wave-fronts leaving an object C, and passing through a rectangular piece of glass DE, and constituting a beam. Every point in a wave-front moves with equal velocity as long as it traverses the same medium; but the point a of a given wave ab enters the glass first, and its velocity is impeded, while the point b retains its original velocity; so that, while the point a moves to a', b moves to b', and the result is that the wave-front assumes a new direction (very much in the same manner as a line of soldiers execute a wheel), and a ray or a line drawn perpendicularly through the series of waves is turned out of its original direction on entering the glass. Again, the extremity c of a given wave-front cd first emerges from the glass, when its velocity is immediately quickened; so that, while d advances to d', c advances to c', and the direction of the ray is again changed. The direction of the ray, after emerging from the glass, is parallel to its direction before entering it, but it has suffered a lateral displacement. Let C represent a section of the wire used in Experiment 262, and the cause of the phenomenon observed will be apparent. If the beam strike the glass perpendicularly, all points of the wave will be checked at the same instant on entering the glass; consequently it will suffer no refraction.

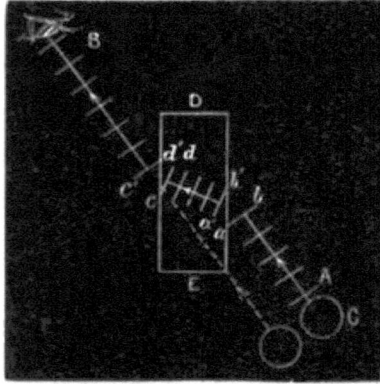

Fig. 265.

280. Index of Refraction. — The deviation of light-waves, in passing from one medium to another, varies with the medium and with the angle of incidence. It

REFRACTION.

diminishes as the angle of incidence diminishes, and is zero when the incident ray is normal (*i.e.* perpendicular to the surface of the medium). It is highly important, knowing the angle of incidence, to be able to determine the direction which a ray will take on entering a new medium. Describe a circle around the point of incidence A (Fig. 266) as a center; through the same point draw IH perpendicular to

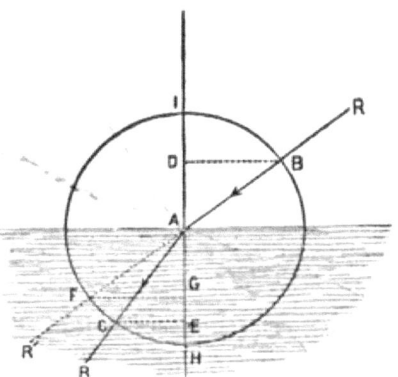

Fig. 266.

the surfaces of the two media, and to this line drop perpendiculars BD and CE from the points where the circle cuts the ray in the two media. Then suppose that the perpendicular BD is $\frac{8}{10}$ of the radius AB; now this fraction $\frac{8}{10}$ is called (in trigonometry) the *sine* of the angle DAB. Hence, $\frac{8}{10}$ is the *sine of the angle of incidence*. Again, if we suppose that the perpendicular CE is $\frac{6}{10}$ of the radius, then the fraction $\frac{6}{10}$ is the *sine of the angle of refraction*. The sines of the two angles are to one another as $\frac{8}{10} : \frac{6}{10}$, or as $4 : 3$. The quotient (in this case $\frac{4}{3} = 1.33+$) obtained by dividing the sine of the angle of incidence by the sine of the angle of refraction is called the *index of refraction*. It can be proved to be the *ratio of the velocity of the incident to that of the refracted light-waves*. It is found that *for the same media the index of refraction is a constant quantity*; *i.e.* the incident ray might be more or less oblique, still the quotient would be the same.

281. Indices of Refraction. — The index of refraction for light-waves in passing from air into water is approximately $\frac{4}{3}$, and from air into glass $\frac{2}{3}$; of course, if the order is reversed, the reciprocal of these fractions must be taken as the indices; *e.g.* from water into air, the index is $\frac{3}{4}$; from glass into air, $\frac{2}{3}$. When a ray passes from a vacuum into a medium, the refractive index is greater than unity, and is called the *absolute index of refraction*. *The relative index of refraction, from any medium A into another B, is found by dividing the absolute index of B by the absolute index of A.*

The refractive index varies with wave-length. The following table is intended to represent *mean indices:* —

TABLE OF ABSOLUTE INDICES.

Air at 0° C., and 760mm pressure . 1.000294	Carbon bisulphide 1.641
Pure water 1.33	Crown glass (about) 1.53
Alcohol 1.37	Flint glass (about) 1.61
Spirits of turpentine 1.48	Diamond (about) 2.5
Humors of the eye (about) . . . 1.35	Lead chromate 2.97

282. Critical Angle; Total Reflection. — Let SS' (Fig. 267) represent the boundary surface between two media, and AO and BO incident rays in the more refractive medium (*e.g.* glass); then OD and OE may represent the same rays respectively after they enter the less refractive

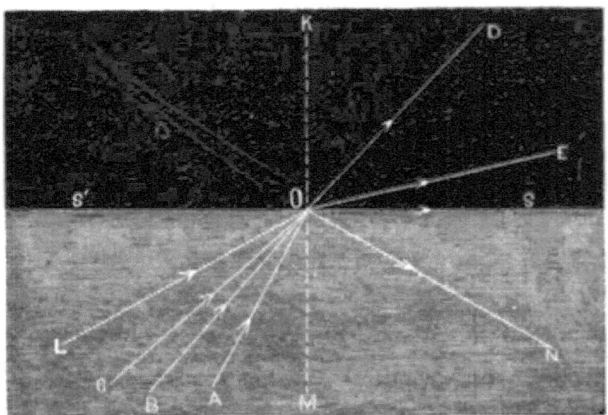

Fig. 267.

medium (*e.g.* air). It will be seen that, as the angle of incidence is increased, the refracted ray rapidly approaches the surface OS. Now, there must be an angle of incidence (*e.g.* COM) such that the angle of refraction will be 90°;

in this case the incident ray CO, after refraction, will just graze the surface OS. This is called *the critical* or *limiting angle*. Any incident ray, as LO, making a larger angle with the normal than the critical angle, cannot emerge from the medium, and consequently is not refracted. Experiment shows that all such rays undergo internal reflection; *e.g.* the ray LO is reflected in the direction ON. Reflection in this case is perfect, and hence is called *total reflection*. *Total reflection occurs when rays in the more refractive medium are incident at an angle greater than the critical angle.*

Surfaces of transparent media, under these circumstances, constitute the best mirrors possible. The critical angle diminishes as the refractive index increases. For water it is about 48½°; for flint glass, 38° 41′; and for the diamond, 23° 41′. Light-waves cannot, therefore, pass out of water into air with a greater angle of incidence than 48½°. The brilliancy of gems, particularly the diamond, is due in part to their extraordinary power of internal reflection, arising from their large indices of refraction.

283. Illustrations of Refraction and Total Reflection.

Experiment 247.— Observe the image of a candle flame reflected by the surface of water in a glass beaker, as in Figure 268.

Experiment 248.— Thrust the closed end of a glass test-tube (Fig. 269) into water, and incline the tube. Look down upon the immersed part of the tube, and its upper surface will look like bur-

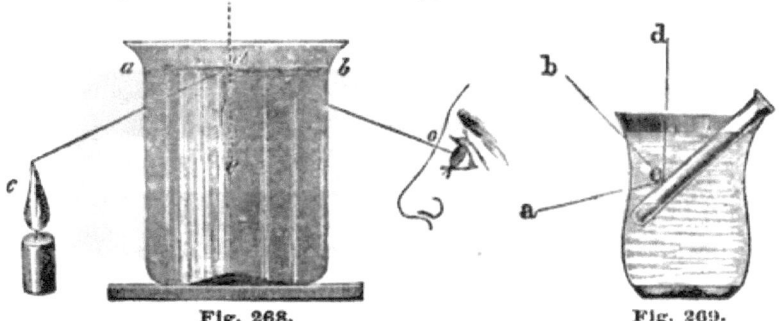

Fig. 268.   Fig. 269.

nished silver, or as if the tube contained mercury. Fill the test-tube with water, and immerse as before; the total reflection which before occurred at the surface of the air in the submerged tube now disappears. Explain.

Section V.

DOUBLE REFRACTION.

284. Double Refraction.

Experiment 249.—Through a card make a pin-hole, and hold the card so that you may see the sky through the hole. Now bring a crystal of Iceland spar (Fig. 270) between the eye and the card, and look at the hole through two parallel surfaces of the crystal. There will appear to be two holes, with light-waves passing through each. Cause the crystal to rotate in a plane parallel with the card, and one of the holes will appear to remain nearly at rest, while the other rotates around the first. A ray *na* immediately on entering the crystal is divided into two parts, one of which obeys the regular law of refraction; the other does not. The former is called the *ordinary ray*; the latter, the *extraordinary ray*. The rays issue from the crystal parallel with each other.

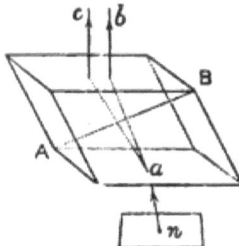

Fig. 270.

In every direction in which one looks through the crystal, except that parallel to AB, objects seen through it appear double. (See Figure 271.) The line AB is called the *optic*

Fig. 271.

axis of the crystal, and is a line around which the molecules of the crystal appear to be arranged symmetrically. A crystal is called *uniaxial* when it has only one optic axis, and *biaxial* when it has two such axes. By far the larger number of crystals of other substances possess the property of causing objects seen through them to appear double. This phenomenon is called *double refraction*.

Section VI.

PRISMS AND LENSES.

285. Optical Prisms. — An optical prism is a transparent, wedge-shaped body. Figure 272 represents a transverse section of such a prism. Let AB be a ray incident upon one of its surfaces. On entering the prism it is refracted *toward* the normal, and takes the direction BC. On emerging from the prism it is again refracted, but now *from* the normal in the direction CD. The object that emits the ray will appear to be at F. Observe that the ray AB, at both refractions, is bent toward the thicker part, or base, of the prism.

Fig. 272.

286. Lenses. — Any transparent medium bounded by two spherical surfaces, or by one plane and the other curved, is a lens.

Experiment 250. — Procure a couple of lenses thicker in the middle than at the edge: strong spectacle glasses, or the large lenses in an opera glass, will answer. Hold one of the lenses in the sun's rays, and notice the path of the beam in dusty air (made so by striking together two blackboard rubbers), after it passes through the lens;

also, that on a paper screen all the rays may be brought to a small circle, or even to a point, not far from the lens. This point is called the *focus*, and its distance from the lens, the *focal length* of the lens.

Find the focal length of this lens, then of the second, and then of the two together. You find the focal length of the two combined is less than of either alone, and learn that the more powerful a lens or combination of them is, the shorter the focal length; that is, the more quickly are the parallel rays that enter different parts of the lens brought to cross one another.

Experiment 251. — Procure a lens thinner in the middle than at its edge. One of the small lenses or eye-glasses of an opera glass will answer. Repeat the above experiment with this lens, and notice that the rays emerging from the lens, instead of coming to a point, become spread out.

Lenses are of two classes, converging and diverging, according as they collect rays or cause them to diverge. Each class comprises three kinds (Fig. 273): —

Class I.		Class II.	
1. Double-convex 2. Plano-convex 3. Concavo-convex. (or meniscus)	Converging, or convex lenses thicker in the middle than at the edges.	4. Double-concave 5. Plano-concave 6. Convexo-concave	Diverging, or concave lenses thinner in the middle than at the edges.

A straight line, as AB, normal to both surfaces of a lens, and passing through its center of curvature, is called its *principal axis*. In every thin lens there is a point in the principal axis called the *optical center*. Every ray

Fig. 273.

that passes through it has parallel directions at incidence and emergence, *i.e.* can suffer at most only a slight lateral displacement. In lenses 1 and 4 it is half-way between their respective curved surfaces. A ray, drawn through the optical center from any point of an object, as A*a* (Fig. 282), is called the *secondary axis* of this point.

287. Effect of Lenses. — We may, for convenience of illustration, regard a convex lens as composed, approximately, of two prisms placed base to base, as A (Fig. 274), and a concave lens as composed of two prisms with their edges in contact, as B. Inasmuch as a beam ordinarily strikes a lens in such a manner that it is bent toward the thicker parts or bases of these approximate prisms, it is obvious that the lens A tends to bend the transmitted rays toward one another, while the lens B tends to separate them. *The general effect of all convex lenses is to converge transmitted rays; that of concave lenses, to cause them to diverge.* Incident rays parallel with the principal axis of a convex lens are brought to

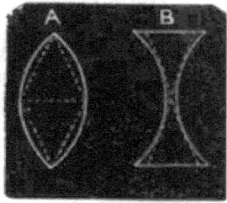

Fig. 274.

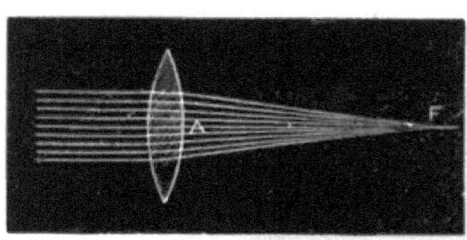

Fig. 275.

a focus F (Fig. 275) at a point in the principal axis. This point is called the *principal focus, i.e.* it is the focus of incident rays parallel with the principal axis. It may be found by holding the lens so that the rays of the sun may fall perpendicularly upon it, and then moving a sheet of paper back and forth behind it until the image of the sun formed on the paper is brightest and smallest. Or, in a room, it may be found approximately, by holding a lens at a considerable distance from a window, regulating the distance so that a distinct image of the window will be projected upon the opposite wall, as in Figure 276. The focal length is the distance of the optical center of the lens

Fig. 276.

to the center of the image on the paper. The shorter this distance the greater is the power of the lens.

If the paper is kept at the principal focus for a short time, it will take fire. The reason is apparent why convex lenses are sometimes called "burning glasses." A pencil of rays emitted from the principal focus F (Fig. 275), as a luminous point, becomes parallel on

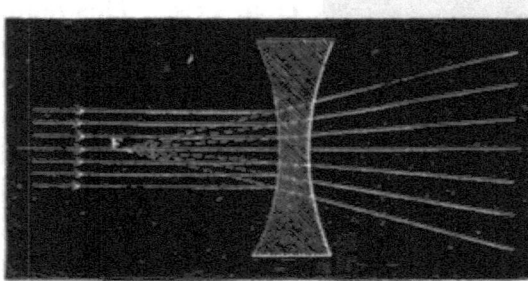

Fig. 277.

emerging from a convex lens. If the rays emanate from a point nearer the lens, they diverge after egress, but the divergence is less than before; if from a point beyond the principal focus, the rays are rendered convergent. A concave lens causes parallel incident rays to diverge as if they came from a point, as F (Fig. 277). This point is therefore its principal focus. It is, of course, a *virtual focus*.

288. Conjugate Foci. — When a luminous point S (Fig. 278) sends rays to a convex lens, the emergent rays converge to another point S'; rays sent

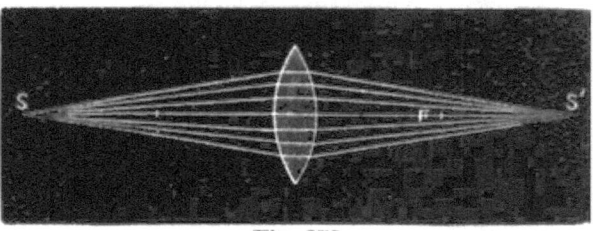

Fig. 278.

from S' to the lens would converge to S. Two points thus related are called *conjugate foci*. The fact that rays which emanate from one point are caused by convex lenses to collect at one point, gives rise to real images, as in the case of concave mirrors.

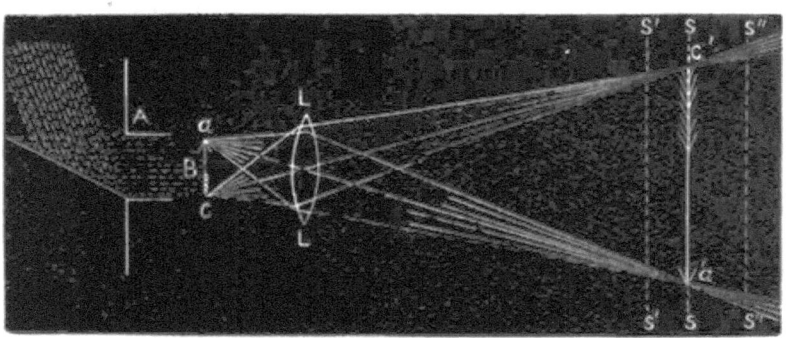

Fig. 279.

289. Images Formed. — Fairly distinct images of objects may be formed through *very small* apertures (page 287); but owing to the small amount of radiant energy that passes through the aperture, the images are very deficient in brilliancy. If the aperture is enlarged, brilliancy is increased at the expense of distinctness. *A convex lens enables us to obtain both brilliancy and distinctness at the same time.*

Experiment 252. — By means of a *porte lumière* A (Fig. 279) introduce a horizontal beam into a darkened room. In its path place

Fig. 280.

some object, as B, painted in transparent colors or photographed on glass. (Transparent pictures are cheaply prepared by photographers for sun-light and lime-light projections.) Beyond the object place a convex lens L (such as represented in Figure 279), and beyond the lens a screen S. The object being illuminated by the beam, all the rays diverging from any point a are bent by the lens so as to come together at the point a'. In like manner, all the rays proceeding from c are brought to the same point c'; and so also for all intermediate points. Thus, out of the innumerable rays emanating from each of the innumerable points on the object, those that reach the lens are guided by it, each to its own appropriate point in the image. It is evident that there must result an image, both bright and distinct, provided the screen is suitably placed, *i.e.* at the place where the rays meet. But if the screen is placed at S' or S'', it is evident that a blurred image will be formed. Instead of moving the screen back and forth, in order to "focus" the rays properly, it is customary to move the lens.

Experiment 253. — Make a series of experiments similar to those (Experiment 237) with the concave mirror. Ascertain the focal length of the convex lens. Place the lens a distance from a white wall about equal to its focal length. Place a candle flame (better the flame of a fish-tail burner) at such a distance the other side of the lens that it will produce a distinct and well-defined image on the wall (Fig. 281). (1) Observe and note on paper the size and kind of image. Advance the flame toward the lens, regulating at the same time the distance between the lens and wall, so as to preserve a distinctness of image. (2) Note the changes which the image undergoes. (3) When the image and flame become of the same size, measure and note the distances of each from the lens. (4) Advance the flame still nearer, and note the changes in the image, until it is impossible to obtain an image on the wall. Measure the distance of the flame from the lens, and compare this distance with the focal length of the lens. (5) Move the flame still nearer. Note whether the rays, after emerging from the lens, are divergent or convergent. (6) See whether an image and

Fig. 281.

an object may change places. (7) Form images of the flame on the wall at different distances from the lens; measure the distances, also the linear dimensions (*e.g.* the width, or the vertical hight) of the images, and determine whether *the linear dimensions of images are proportional to their distances from the lens.*

290. To Construct the Image Formed by a Convex Lens. — Given the lens L (Fig. 282), whose principal focus is at F (or F', for rays coming from the other direction), and object AB in front of it; any two of the many rays from A will determine where

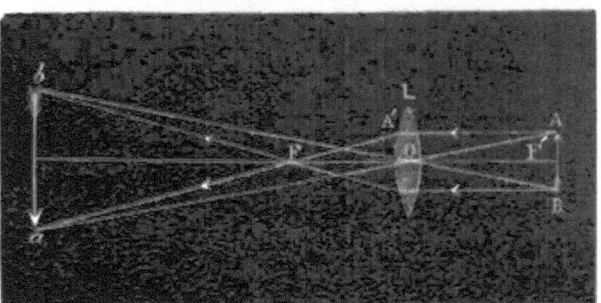

Fig. 282.

its image *a* is formed. The two that can be traced easily are, the one along the secondary axis AO*a*, and the one parallel to the principal axis AA': the latter will be deviated so as to pass through the principal focus F, and will afterward intersect the principal axis at some point a; so this is the conjugate focus of A; similarly for B, and all intermediate points along the arrow. Thus, a *real inverted image* is formed at ab.

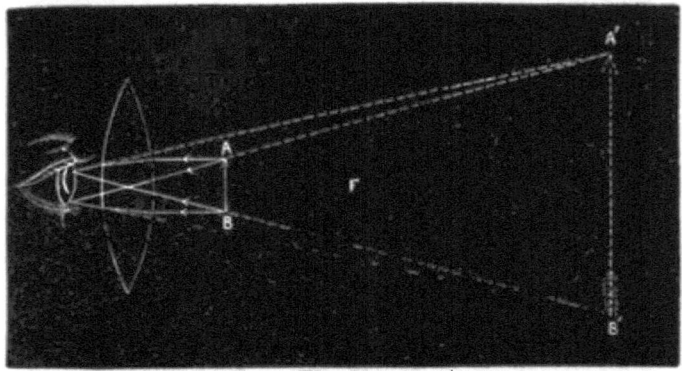

Fig. 283.

291. Virtual Images; Simple Microscope. — Since rays that emanate from a point nearer the lens than the principal focus diverge after egress, it is evident that their focus must be virtual and on the same side of the lens as the object. Hence, *the image of an object placed nearer the lens than the principal focus is virtual, magnified, and erect*, as shown in Figure 283. A convex lens used in this

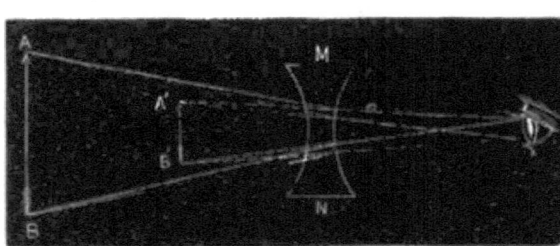

Fig. 284.

manner is called a *simple microscope*.

Since the effect of concave lenses is to scatter transmitted rays, pencils of rays emitted from A and B (Fig. 284), after refraction, diverge as if they came from A' and B', and the image will appear to be at A B'. Hence, *images formed by concave lenses are virtual, erect, and smaller than the object*.

PRISMS AND LENSES. 313

292. Spherical Aberration. — In all ordinary convex lenses the curved surfaces are spherical, and the angles which incident rays make with the little plane surfaces, of which we may imagine the spherical surface to be made

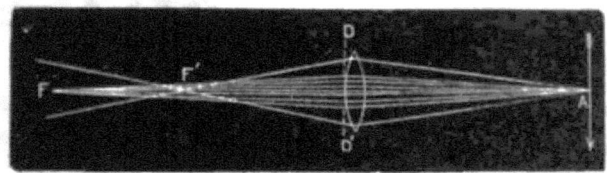

Fig. 285.

up, increase rapidly toward the edge of the lens. Thus, while those rays from a given point of an object, as A (Fig. 285), which pass through the central portion, meet approximately at the same point F, those which pass through the marginal portion are deviated so much that they cross the axis at nearer points, *e.g.* at F'; so a blurred image results. This wandering of the rays from a single focus is called *spherical aberration*. The evil may be largely corrected by interposing a diaphragm DD', provided with a central aperture, smaller than the lens, so as to obstruct those rays that pass through the marginal part of the lens.

Experiment 254. — (Illustrating spherical aberration.) Cut a cardboard disk as large as the convex lens (Fig. 280). Cut a ring of holes near the circumference, and also a ring near the center. Support the disk close to the lens, so as to cover one of its surfaces. Place the whole in a beam from a *porte lumière*. Catch refracted beams on a screen. Move the screen away from the lens. The beams through the outer ring of spots are the first to cross one another and form an image. Further away, the inner beams coincide, forming an image. The outer ones having crossed, form a ring of spots.

Section VII.

PRISMATIC ANALYSIS OF LIGHT-WAVES. — SPECTRA.

293. Analysis of Light-Waves which Produce the Sensation of White.

Experiment 255. — Place the disk with adjustable slit in the aperture of a *porte lumière*, so as to exclude all light-waves from a darkened room except those which pass through the slit. Near the slit interpose a double-convex lens of (say) 10-inch focus. A narrow sheet of light will traverse the room and produce an image AB (Fig. 286) of the slit on a white screen placed in its path. Now place a glass prism C in the path of the narrow sheet of light-waves and near to the lens with its edge vertical. (1) The light-waves now are not only turned

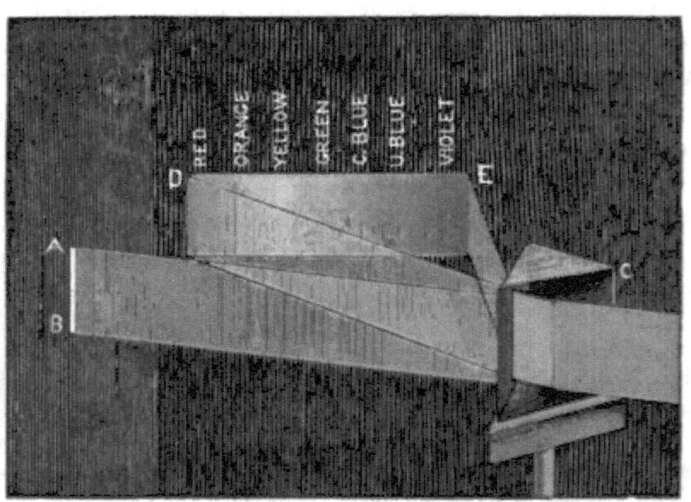

Fig. 286.

from their former path, but that which before was a narrow sheet, is, after emerging from the prism, spread out fan-like into a wedge-shaped body, with its thickest part resting on the screen. (2) The image, before only a narrow, vertical band, is now drawn out into a long horizontal ribbon, DE. (3) The image, before white, now presents all the colors of the rainbow, from red at one end to violet at the other; it passes gradually through all the gradations of red, orange, yellow, green, blue, and violet. (The difference in deviation between the red and the violet is purposely much exaggerated in the figure.)

From this experiment we learn (1) that *white waves* (*i.e.* those waves which are capable of producing the sensation of white) *are not simple* in their composition, *but the result of a mixture.* (2) *The color waves of which white waves are composed may be separated by refraction.* (3) *The cause of the separation is due to the different degrees of deviation which they undergo by refraction.* Red waves, which are always least turned aside from a straight path, are the least refrangible. Then follow orange, yellow, green, blue, and violet in the order of their refrangibility. The many-colored ribbon DE is called the *solar spectrum*. This separation of white waves into their constituents is called *dispersion*. The variety of color waves of which white waves are composed is really infinite; but we name the seven principal ones as follows: *red, orange* (*or citron*), *yellow, green, cyan-blue, ultramarine-blue, and violet;* these are called the *prismatic colors*. The names of the blues are derived from the names of the pigments which most closely resemble them.

294. The Rainbow. — The rainbow is an illustration of a solar spectrum on a grand scale. It is the result of refraction, reflection, and dispersion of sunlight by falling raindrops. Let spheres 1 and 2 (Fig. 287) represent drops at the extreme opposite edges of the bow. The eye is in a position to receive after the dispersion and internal reflection of the light-waves within this drop, only the red waves; consequently this part of the bow appears red. So, likewise, from drop 2, the eye receives only violet; consequently this edge appears violet. In like manner, the intermediate colors of the bow are sifted out.

Outside the primary bow a *secondary bow* (Fig. 288) is sometimes seen. Drops 3 and 4 (Fig. 287) are supposed to be at the opposite edges of the

316 RADIANT ENERGY.

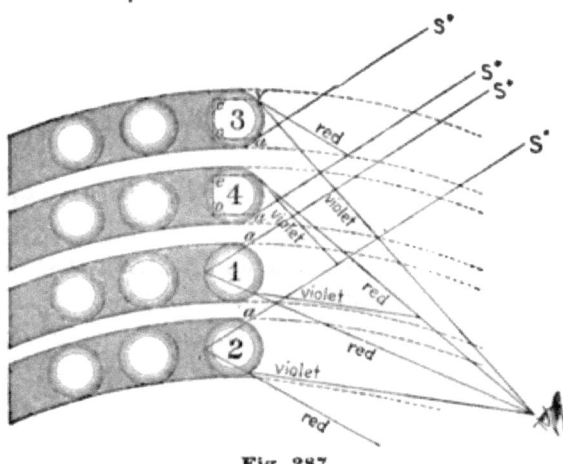

Fig. 287.

secondary bow. It will be seen that the light-waves undergo two internal reflections within the drops which produce this bow. The colors of this bow are in reverse order of those of the primary bow, and less brilliant.

Fig. 288.

295. Synthesis of White Waves. — The composition of white waves has been ascertained by the process of analysis; can it be verified by *synthesis?* — *i.e.* can the colors after dispersion be reunited? and, if so, will white be restored?

Experiment 256. — Place a second prism (2) in such a position ($\triangle\nabla$) that light-waves which have passed through one prism (1), and been refracted and decomposed, may be refracted back, and the colors will be reblended, and a white image of the slit will be restored on the screen.

Experiment 257. — Place a large convex lens, or a concave mirror, so as to receive the colors after dispersion by a prism, and bring the rays to a focus on a screen. The image produced will be white.

296. Cause of Color Revealed by Dispersion. — *Color is determined solely by the number of waves emitted by a luminous body in a second of time, or by the corresponding wave-length.* In a dense medium, the short waves are more retarded than the longer ones; hence they are more refracted. This is the cause of dispersion. The ether waves diminish in length from the red to the violet. As pitch depends on the number of aërial waves which strike the ear in a second, so color depends on the number of ethereal waves which strike the eye in a second.

From well-established data, determined by a variety of methods (see larger works), physicists have calculated the number of waves that succeed one another for each of the several prismatic colors, and the corresponding wave-lengths; the following table contains the results. The letters A, C, D, etc., refer to Fraunhofer's lines (see Plate I.).

		Length of waves in millimeters.	Number of waves per second.
Dark red	A	.000760	395,000,000,000,000
Orange	C	.000656	458,000,000,000,000
Yellow	D	.000589	510,000,000,000,000
Green	E	.000527	570,000,000,000,000
C. Blue	F	.000486	618,000,000,000,000
U. Blue	G	.000431	697,000,000,000,000
Violet	H	.000397	760,000,000,000,000

There is a limit to the sensibility of the eye as well as of the ear. The limit in the number of vibrations appreciable by the eye lies approximately within the range of numbers given in the above table; *i.e.* if the succession of waves is much more or less rapid than indicated by these numbers, they do not produce the sensation of sight.

297. Continuous Spectra. — *All luminous solids and liquids give continuous spectra.* If the spectrum is not complete, as when the temperature is too low, it will begin with red, and be continuous as far as it goes.

298. Spectroscope. — A small instrument called a pocket *spectroscope*[1] will answer for all experiments given in this book. More elaborate experiments require more elaborate apparatus, a description of which must be sought for in larger works on this subject. This instrument contains three or more prisms, A, B, and C (Fig. 289). The prisms are enclosed in a brass tube D, and this tube in another tube E. F is a convex lens, and G is an adjustable slit. By moving the inner tube back and forth, the instrument may be so focused that parallel rays will fall upon

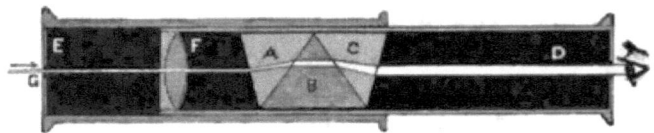

Fig. 289.

prism A. By varying the kind of glass used in the different prisms,[2] as well as their structure, the deviation of light-waves from a straight path, in passing through them, is overcome, while the dispersion is preserved. On account of the directness of the path of light-waves through it, this instrument is called a *direct-vision spectroscope*.

299. Bright Line, Absorption, or Reversed Spectra.
Experiment 258. — Open the slit about one-sixteenth of an inch wide, by turning the milled ring M (Fig. 290), and look through the spectroscope at the sky (not at the sun, for its light-waves are too intense for the eye), and you will see a continuous spectrum.

Fig. 290.

---

[1] It is expected that the pupil will be provided with a pocket spectroscope, the cost of which need not exceed ten dollars.

[2] A and C are crown-glass, and B is flint-glass.

Experiment 259. — Repeat the last experiment with a candle, kerosene, or ordinary gas flame, and you will obtain similar results.

Experiment 260. — Take a piece of platinum wire 16 inches long. Seal one end by fusion to a short glass tube for a handle. Bend the wire at a right angle. Dip a portion of the wire into a strong solution of common salt, and support it by a clamp in the midst of the almost invisible and colorless flame of a Bunsen burner or alcohol lamp (Fig. 291). Instantly the flame becomes luminous and colored a deep yellow. Examine it with a spectroscope, and you will find, instead of a continuous spectrum beginning with red, only a bright, narrow line of yellow, in the yellow part of the spectrum, next the orange. Your spectrum consists essentially of a single bright yellow line on a comparatively dark ground (see Sodium, Plate I., frontispiece).

Fig. 291.

Experiment 261. — Heat the platinum wire until it ceases to color the flame, then dip it into a solution of chloride of lithium, and repeat the last experiment. You obtain a carmine-tinted flame, and see through the spectroscope a bright red line and a faint orange line (see Lithium, Plate I.).

Experiment 262. — Use potassium hydrate, and you obtain a violet-colored flame, and a spectrum consisting of a red line and a violet line (the latter is very difficult to see even with the best instruments). Use strontium nitrate, and obtain a crimson flame, and a spectrum consisting of several lines in the red and the orange, and a blue line (see Potassium and Strontium, Plate I.).

Experiment 263. — Use a mixture of several of the above chemicals, and you will obtain a spectrum containing all the lines that characterize the several substances.

Every chemical compound used in the above experiments contains a different metal, *e.g.* common salt contains the metal sodium; the other substances used successively contain respectively the metals lithium, potassium, and strontium. These metals, when introduced into the flame, are vaporized, and we get their spectra when in a gaseous state. *All incandescent gases, unless under great pressure, give discontinuous, or bright line, spectra, and no two gases give the same spectra.*

300. Dark-line Spectra.

Experiment 264. — Close the slit of the spectroscope so that the aperture will be very narrow; direct it once more to the sky, and slowly move the inner tube back and forth, and you will find, with a certain suitable adjustment which may be obtained by patient trial, that the solar spectrum is not in reality continuous, but is crossed by several *dark lines* (see Solar Spectrum, Plate I.).

Remark. — In general it is best to focus either the D line in the orange, or the E line in the green. The inner sliding tube ought to be drawn out a little when examining the blue end of the spectrum, and pushed in for focusing the lines in the red.

Experiment 265. — Put a few copper turnings in a test-tube, add a little nitric acid. Hold the tube causing the colored vapor before the slit, and notice the black bands.

Experiment 266. — The electric light is now in so common use that it may be possible to perform this experiment. Between the electric light and the spectroscope introduce the flame of a Bunsen burner, and color it yellow with salt. Examine the spectrum formed through this yellow flame.

In the last experiment you would naturally expect to find the yellow part of the spectrum uncommonly bright, for there would apparently be added to the yellow waves of the electric light the yellow waves of the salted flame. But precisely where you would look for the brightest yellow, there you discover that the spectrum is crossed by a dark line. If you use salts of lithium, potassium, and strontium in a similar manner, you will find in every case your spectrum crossed by dark lines where you would expect to find bright lines. Remove the Bunsen flame, and the dark lines disappear. It thus appears that *the vapors of different substances absorb or quench the very same waves that they are capable of emitting;* very much, it would seem, as a given tuning-fork selects from various sound-waves only those of a definite length corresponding to its own vibration-period. The dark places of the spectrum are illuminated by the salted flame; but these places are so feebly illuminated in comparison with those places illuminated by the electric light, that the former appear dark by contrast. Light-waves transmitted through certain liquids (as sulphate of quinine and blood) and certain solids (as some colored glasses) produce *dark-line* spectra. These spectra are obtained only when light-waves pass through media capable of absorbing waves of certain length; hence they are commonly called *absorption spectra*. Since a given vapor causes dark lines precisely where, if it were itself the only radiator of light-waves, it would cause bright lines, dark-line spectra are frequently called *reversed spectra*. There are then three kinds of spectra: *continuous spectra*, produced by luminous solids, liquids, or, as has been found in a few instances, gases under great pressure; *bright-line spectra*, produced by luminous vapors; and *absorption spectra*, produced by light-waves that have been sifted by certain media.

301. Spectrum Analysis. — More elaborate spectroscopes contain many prisms, by which the *purity* of the spectrum is greatly increased. (By purity is meant a freedom from the overlapping of images of the slit, by which many lines of the spectrum are obscured.) They also contain an illuminated scale which may be seen adjacent to the spectrum, by which the exact position of the lines and their relative distances from one another can be accurately determined, and a telescope by which the spectrum and scale may be magnified. The positions of some of the prominent lines of the solar spectrum were first determined, mapped, and distinguished from one another by certain letters of the alphabet, by Fraunhofer; hence the dark lines of the solar spectrum are commonly called *Fraunhofer's lines*. So far as discovered, no two substances have a spectrum consisting of the same combination of lines; and, in general, different substances but very rarely possess lines appearing to be common to both. Hence, when we have once observed and mapped the spectrum of any substance, we may ever after be able to recognize the presence of that substance when emitting light-waves, whether it is in our laboratory or in a distant heavenly body.

The spectroscope, therefore, furnishes us a most efficient means of detecting the presence (or absence) of any elementary substance, even when it is combined or mixed with other substances. It is not necessary that the given substance should exist in large quantities; for example, a fourteen-millionth of a milligram of sodium can be detected by the spectroscope.

302. Celestial Chemistry and Physics. — The spectrum of iron has been mapped to the extent of 460 bright lines. The solar spectrum furnishes dark lines corresponding to nearly all these bright lines. Can there be any doubt of the existence of iron in the sun? By examination of the reversed spectrum of the sun, we are able to determine with certainty the existence there of sodium, calcium, copper, zinc, magnesium, hydrogen, and many other known substances. The moon and other heavenly bodies that are visible only by reflected sunlight give the same spectra as the sun, while those that are self-luminous give spectra which differ from the solar spectrum.

303. Relative Heating and Chemical Effects of Ether-Waves of Different Lengths. — If a sensitive thermometer is placed in different parts of the solar spectrum, it will indicate heat in all parts; but the heat generally increases from the violet toward the red. It does not cease, however, with the limit of the visible spectrum; indeed, if the prism is made of flint glass, the greatest heat is just beyond the red. A strip of paper wet with a solution of chloride of silver suffers no change in the dark; in the light-waves it quickly turns black; exposed to the light-waves of the solar spectrum, it turns dark, but quite unevenly. The change is slowest in the red, and constantly increases, till about the region indicated by G (see Solar Spectrum, Plate I.), where it attains its maximum; from this point it falls off, and ceases at a point considerably beyond the limit of the violet. It thus appears that the solar spectrum is not limited to the visible spectrum, but extends beyond at each extremity. Those waves that are beyond the red are usually called the *infra-red* waves, while those that are beyond the violet are called the *ultra-violet* waves. The infra-red waves are of longer vibration-period, and the ultra-violet of shorter period, than the light-waves.

304. Only one Kind of Radiation. — It has been shown that radiant energy may produce three distinct effects, according to the means by which it is absorbed or the sense which it affects. But the radiant energy producing these three and other effects is but one and the same thing. The only difference in radiant energy is that which is common to all wave-motion, viz. *difference in wave-length and difference in amplitude*, the latter causing the wave to possess more energy as the amplitude is greater. By a lamp-blacked surface nearly all the radiant energy of waves of whatever length is absorbed and transformed into heat. By exposing such a surface to spectra we learn that the longer waves possess more energy than the shorter. On the other hand, most chemical mixtures which are affected by sunlight are more sensitive to the shorter waves, *i.e.* this rate of vibration stimulates chemical action to a greater extent. But the sense of sight is affected only by waves within the range already stated, § 297.

While waves traverse the ether there is neither heat nor light (*i.e.* sensation); hence the propriety of applying either of these terms to a train of waves traversing the ether may well be called in question. Yet this is all that traverses the space between the sun and the earth.

305. Chromatic Aberration. — There is a serious defect in ordinary convex lenses, to which we have not before alluded, called *chromatic aberration*, which has required the highest skill to correct. The convex lens both *refracts* and *disperses* the light-waves that pass through it. The tendency, of course, is to bring the more refrangible rays, as the violet, to a focus much sooner than the less refrangible rays, such as the red. The result is a disagreeable coloration of the images that are formed by the lens, especially by that portion of the light-waves that passes through the lens near its edges. This evil has been overcome very effectually by combining with the convex lens a plano-concave lens. Now, if a crown-glass convex lens is taken, a flint-glass concave lens may be prepared that will correct the dispersion of the former without neutralizing all its refraction.[1] A compound lens, composed of these two lenses (Fig. 292) cemented together, constitutes what is called an *achromatic lens*.

Fig. 292.

---

[1] The refractive and dispersive powers of the two lenses are not proportional.

Section VIII.

COLOR.

306. Color Produced by Absorption. — "Color is a sensation" [Alfred Daniell]. "All objects are black in the dark"; this is equivalent to saying that *without light-waves there is no color*. Is color due to some quality of an object, or is it due to a quality of the light-waves which illuminate the object?

Experiment 267. — We have found that common salt introduced into a Bunsen flame renders it luminous, and that the light-waves, when analyzed with a prism, is found to contain only yellow. Expose papers or fabrics of various colors to these light-waves in a darkened room. *No one of them exhibits its natural color, except yellow.*

Experiment 268. — Hold a narrow strip of red paper or ribbon in the red portion of the solar spectrum; it appears red. Slowly move it toward the other end of the spectrum; on leaving the red it becomes darker, and when it reaches the green it is quite black, or colorless, and remains so as it passes the other colors of the spectrum. Repeat the experiment, using other colors, and notice that only in light-waves of its own color does each strip of paper appear of its color; while in all other colors it is dark.

These experiments show that (1) *color is due to a quality of the light-waves which illuminate, and not of the object illuminated, though by a conventionality of language we ascribe colors to objects;* (2) *in order that an object may appear of a certain color, it must receive light-waves of that color; and of course if it receives other color waves at the same time, it must be capable of absorbing or transmitting them.* The energy of the waves absorbed is converted into heat, and warms the object. When white waves (*i.e.* those capable of producing the sensation of white) strike an object, it appears white if it reflects all the color waves. If red waves fall upon the same object, it appears red, for it is capable of reflecting red waves; or it appears green, if green waves alone fall on it. If white waves fall upon an object, and all the color waves are absorbed except the blue, the object appears blue. When we paint our houses we do not apply color to them. We apply substances, called *pigments*, that have a property of absorbing all the color waves except those which we would have our houses appear.

Experiment 269. — By means of a *porte lumière* introduce a beam into a dark room. Cover the orifice with a deep red (copper) glass. The white waves, in passing through the glass, appear to be colored red. *Does the glass color the waves red?*

Experiment 270. — With the slit, lens, and prism form a solar spectrum, and between the prism and screen interpose the red glass. Very few light-waves, except the red, are transmitted; the rest are absorbed by the glass.

It thus appears that a red transparent body is red because *it transmits few light-waves except the red*, not because the body colors the waves.

307. Sky Colors.

Experiment 271. — Dissolve a little white castile soap in a tumbler of water; or, better, stir into the water a few drops of an alcoholic solution of mastic, enough to render the water slightly turbid. Place a black screen behind the tumbler, and examine the liquid by reflected sunlight, — the liquid appears to be blue. Examine the liquid by transmitted sunlight, — it now appears yellowish red.

Sky-light is the result of reflected light-waves. The particles of atmospheric dust (of water, probably) that pervade the atmosphere, like the fine particles of mastic suspended in the water, reflect blue light-waves; while beyond the atmosphere is a black background of darkness. But we must not, from this, conclude that the atmosphere is blue; for, unlike blue glass, but like the turbid liquid, it transmits yellow and red rays freely, so that, seen by reflected light-waves, it is blue, but seen by transmitted light-waves it is yellowish red.

Experiment 272. — Pour some of the turbid liquid into a small test-tube, and examine it and the tumbler of liquid by transmitted light-waves; the former appears almost colorless, while the latter is quite deeply colored.

When the sun is near the horizon, its rays travel a greater distance in the air to reach the earth than when it is in the zenith; consequently, there is a greater loss by absorption and reflection in the former case than in the latter. But the yellow and red rays suffer less destruction, proportionally, than the other colors; consequently, these colors predominate in the morning and evening.

The remarkable "yellow days" of the summer of 1882 are explained in this way. The atmosphere on this continent was remarkably turbid during those days.

308. Mixing Colors. — A mixture of all the prismatic colors, in the proportion found in sunlight, produces white. Can white be produced in any other way?

Experiment 273. — On a black surface A (Fig. 293), about 2 inches apart, lay two small rectangular pieces of paper, one yellow

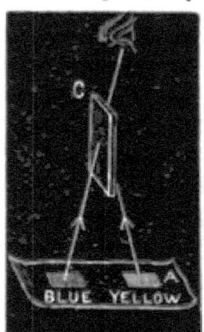

Fig. 293.

and the other blue. In a vertical position between, and from 2 inches to 6 inches above, these papers, hold a slip of plate glass C. Looking obliquely down through the glass you may see the blue paper by transmitted light-waves and the yellow paper by reflection. That is, you see the object itself in the former case, and the image of the object in the latter case. By a little manipulation, the image and the object may be made to overlap one another, when both colors will apparently disappear, and in their place the color which is the result of the mixture will appear. In this case it will be white, or, rather, *gray*, which *is white of a low degree of luminosity*. If the color is yellowish, lower the glass; if bluish, raise it.

Experiment 274. — With the rotating apparatus, rotate the disk (Fig. 294) which contains only yellow and blue. The colors so blend

COLOR. 327

(*i.e.* the sensations) in the eye as to produce the sensation of gray, *i.e.* white of low luminosity.

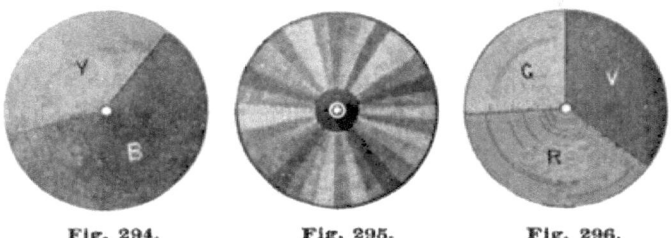

Fig. 294.   Fig. 295.   Fig. 296.

Figure 295 represents "Newton's disk," which contains the seven prismatic colors arranged in a proper proportion to produce gray when rotated.

In like manner, you may produce white by mixing purple and green; or, if any color on the circumference of the circle (see Complementary Colors, Plate I.) is mixed with the color exactly opposite, the resulting color will be white. Again, the three colors, red, green, and violet, arranged as in Figure 296, with rather less surface of the green exposed than of the other colors, will give gray. Green mixed with red, in varying proportions, will produce any of the colors in a straight line between these two colors in the diagram (Plate I.); green mixed with violet will produce any of the colors between them; and violet mixed with red gives purple.

All colors are represented in the spectrum, except the purple hues. The latter form the connecting link between the two ends of the spectrum. Our color chart (Plate I.) is intended to represent the sum total of all the sensations of color. By means of this chart we may determine the result of the (optical) mixture of any two colors as follows: Find the places occupied upon the chart by the two colors which are to be mixed, and unite the two points by a straight line. The color produced by the mixture will invariably be found at the center of this line.

309. Mixing Pigments.

Experiment 275. — Mix a little of the two pigments, chrome yellow and ultramarine blue, and you obtain a green pigment.

The last three experiments show that mixing certain colors, and mixing pigments of the same name, may produce very different results. In the first experiments you mixed colors; in the last experiment you did not mix colors, and we must seek an explanation of the result obtained. If a glass vessel with parallel sides containing a blue solution of sulphate of copper is interposed in the path of the light-waves which form a solar spectrum, it will be found that the red, orange, and yellow waves are cut out of the spectrum, *i.e.* the liquid absorbs these waves. And if a yellow solution of *bichromate* of potash is interposed, the blue and violet waves will be absorbed. It is evident that, if both solutions are interposed, all the colors will be destroyed, except the green, which alone will be transmitted; thus: —

    Cancelled by the blue solution, R̸ Ø Y̸ G B V.
    Cancelled by the yellow solution, R O Y G B̸ V̸.
    Cancelled by both solutions, R̸ Ø Y̸ G B̸ V̸.

In a similar manner, when white waves strike a mixture of yellow and blue pigments on the palette, they penetrate to some depth into the mixture; and, during its passage in and out, all the colors are destroyed, except the green; so the mixed pigments necessarily appear green. But when a mixture of yellow and blue waves enters the eye, we get, as the result of the *combined* sensations produced by the two colors, the sensation of white; hence a mixture of yellow and blue gives white.

The color square 3 (Plate I.) represents the result of the mixture of pigments 1 and 2; while 4 represents the result of the optical mixture of the same colors.

310. Complementary Colors.

Experiment 276. — On a piece of white, or better, gray, paper, lay a circular piece of blue paper 15^{mm} in diameter. Attach one end of a piece of thread to the colored paper, and hold the other end in the hand. Place the eyes within about 15^{cm} of the colored paper, and look steadily at the center of the paper for about fifteen seconds; then, without moving the eyes, suddenly pull the colored paper away, and instantly there will appear on the gray paper an image of the colored paper, but the image will appear to be yellow. This is usually called an *after-image*. If yellow paper is used, the color of the after-image will be blue; and if any other color given in the diagram (Plate I.), the color of its after-image will be the color that stands opposite to it.

This phenomenon is explained as follows: When we look steadily at blue for a time, the eyes become fatigued by this color, and less susceptible to its influence, while they are fully susceptible to the influence of other colors; so that when they are suddenly brought to look at white, which is a compound of yellow and blue, they receive a vivid impression from the former, and a feeble impression from the latter; hence the predominant sensation is yellow. Any two colors which together produce white are said to be *complementary* to each other. The opposite colors in the diagram (Plate I.) are complementary to one another.

311. Effect of Constrast. — When any two colors given in the circle (Plate I.) are brought in contrast, as when they are placed next one another, the effect is to move them farther apart. For example, if red and orange are brought in contrast, the orange assumes more of a yellowish hue, and the red more of a purplish hue. Colors that are already as far apart as possible, *e.g.* yellow and blue, do not change their hue, but merely cause one another to appear more brilliant.

312. Color Produced by Interference.

Experiment 277. — In a vise or other convenient instrument, press two clean pieces of thick plate glass firmly together. A number of colors will be seen arranged in a certain order, and forming curves more or less regular around the point of pressure.

This, together with many other kindred color phenomena, is caused by the mutual destruction by *interference* of certain of the colors which compose white, the resulting colors being the product of the combination of those which are not so extinguished. Much as certain over-tones might destroy one another, and the quality of the resulting sound would be determined by the composition of the surviving tones.

Thin, transparent films of varying thickness, such as the film of a soap bubble, are well suited to show the effects of interference of light-waves. Some of the light-waves which strike the anterior surface of the film are reflected; another portion enters the film, and is reflected from the posterior surface; but, by travelling twice through the film, the waves lose ground, so that, on emergence, their phases may or may not correspond with the phases of the former portion: this will depend evidently upon the thickness of the film at a given point, and the length of the waves striking that point. In this manner the phenomena obtained in the experiment are explained; the film in this case is the layer of air between the two surfaces of glass.

Colors are produced by reflection from the surfaces of thin transparent films of all kinds; for example, the colors of the soap bubble, of oil on water, of the thin coating of metallic oxide formed in tempering steel.

Section IX.

THERMAL EFFECTS OF RADIATION.

313. Diathermancy and Athermancy.

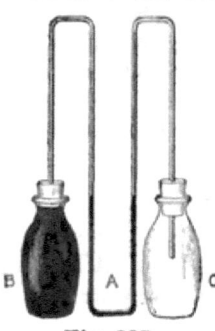

Fig. 297.

Experiment 278. — Prepare a *differential thermometer* with two glass flasks and a glass tube, as represented in Figure 297. Cover one of the flasks with lamp-black by holding it above a smoking kerosene flame. Place colored liquid in the bend A. Stopper both vessels tightly and expose the apparatus to the direct rays of the sun. The rays pass through the clean glass and through the air within, affecting the temperature of either but little. But the lamp-black absorbs the radiations, the flask becomes heated, the enclosed air becomes heated by contact with the heated flask, the heated air expands and pushes the liquid in the tube toward the cooler flask.

What becomes of radiations that strike a body depends largely upon the character of the body. If the nature of the body is such that its molecules can accept the motion of the ether, the undulations of ether are said to be absorbed by the body, and the body is thereby heated; that is, the radiant energy is transformed into heat energy. A good illustration of this is the experiment with blackened glass. On the other hand, the unblackened glass allows the radiations to pass freely through it, and very little is transformed into heat. Notice how cold window-glass may remain, while radiations pour through it and heat objects within the room. It must be constantly borne in mind that *only those radiations that a body absorbs heat it; those that pass through it do not affect its temperature.* Bodies that transmit radiant energy freely are said to be *diathermanous*, while those that absorb it largely are called *athermanous*. The most diathermanous solid is rock salt. Among the most athermanous solids are lamp-black and alum. Carbon bisulphide, among liquids, is exceptionally transparent to all forms of radiation; while water, transparent to short waves, absorbs the longer waves, and is thus quite athermanous.

Dry air is almost perfectly diathermanous. All of the sun's radiations that reach the earth pass through a layer of air from fifty to two hundred miles in depth, which contains a vast amount of aqueous vapor. This vapor, like water, is comparatively opaque to long waves; hence it modifies very much the character of the radiations which reach the earth. This fact enables us to understand the method by which our atmosphere becomes heated. First, a very considerable portion of the radiant energy which comes to us from the sun, in the form of relatively long waves, is stopped by the watery vapor in the air, which is, in consequence, heated. Most of that which escapes this absorption heats the earth by falling upon it. The warmed earth loses its heat, — partly by conduction to the air, still more largely by radiation outward. The form of radiation, however, has been greatly changed; for now, coming from a body at a low temperature, it is chiefly in long waves that the energy is transmitted; while, as we have seen, it was largely in the form of short waves that the earth received its heat. But it is exactly these long waves which are most readily stopped by the atmosphere; hence the atmosphere, or rather the aqueous vapor of the atmosphere, acts as a sort of trap for the energy which comes to us from the sun. Remove the watery vapor (which serves as a "blanket" to the earth) from our atmosphere, and the chill resulting from the rapid escape of heat by radiation would put an end to all animal and vegetable life. Glass does not screen us from the sun's radiations, but it can very effectually screen us from the radiations from a stove or any other terrestrial object. Glass is diathermanous to the sun's radiations (simply because they have already lost most of the very long waves by atmospheric absorption), but quite athermanous to other radiations. This is well illustrated in the case of hot-beds and green-houses. The sun's radiations pass through the glass of these enclosures, almost unobstructed, and heats the earth; but the radiations given out in turn by the earth are such as cannot pass out through the glass; hence the heat is retained within the enclosures.

314. All Bodies Radiate Heat. — Hot bodies *usually* part with their heat much more rapidly by radiation than by all other processes combined. But cold bodies, like ice, radiate heat even when surrounded by warm bodies. This must be so from the nature of the case, for the molecules of the coldest bodies possess some motion, and, being surrounded by ether, they cannot move without imparting some of their motion to the ether, and to that extent become themselves colder.

315. Theory of Exchanges. — Let us suppose that we have two bodies, A and B, at different temperatures, A warmer than B. Radiation takes place not only from A to B, but from B to A; but, in consequence of A's excess of temperature, more radiant energy passes from A to B than from B to A, and this continues until both bodies acquire the same temperature. At this point radiation by no means ceases, but each now gives as much as it receives, and thus equilibrium is kept up. This is known as the "Theory of Exchanges."

316. Good Absorbers, Good Radiators.

Experiment 279. — Select two small tin boxes of equal capacity; one should be bright outside, while the other should be covered thinly with soot from a candle-flame. Cut a hole in the cover of each box large enough to admit the bulb of a thermometer. Fill both boxes with hot water, and introduce into each a thermometer. They will register the same temperature at first. Set both in a cool place, and in half an hour you will find that the thermometer in the blackened box registers several degrees lower than the other. Then fill both with cold water, and set them in front of a fire or in the sunshine, and it will be found that the temperature in the blackened box rises faster.

As bodies differ widely in their absorbing power, so they do in their radiating power, and it is found to be universally true that *good absorbers are good radiators, and bad absorbers are bad radiators*. Much, in both cases, depends upon the character of the surface as well as the substance. Bright, polished surfaces are poor absorbers and poor radiators; while tarnished, dark, and roughened surfaces absorb and radiate heat rapidly. Dark clothing absorbs radiations and radiates more rapidly than light clothing.

Section X.

SOME OPTICAL INSTRUMENTS.

317. Compound Microscope. — The *simple microscope* was described on page 312. When it is desired to magnify an object more than can be done conveniently and with distinctness by a single lens, two convex lenses are used, — one (O, Fig. 298) called the *object-glass*, to form a magnified real image $A'B'$ of the object AB; and the other (E) called the *eye-glass*, to magnify this image so that the image $A'B'$ appears of the size $A''B''$.

Fig. 298.

Hence the *compound microscope* is virtually a simple microscope applied not to the object, but to its image already magnified by the object lens. Both lenses should be achromatic and aplanatic (free from spherical aberration).

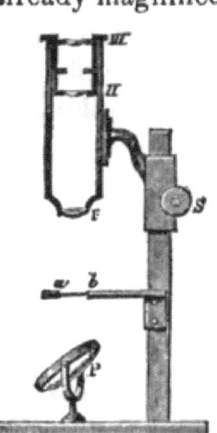

Fig. 299.

The eye-piece is made of two or more lenses, because it is found that if the refractions are thus distributed, the extent of the useful field may be greatly increased. Ordinarily two lenses are sufficient.

The article to be examined is placed on a glass stage, ab (Fig. 299), and, if the object is transparent, it is strongly illuminated by focusing light upon it by means of a concave mirror, P. If the object is opaque, it is illuminated by light-waves converged upon it obliquely from above by a convex lens not shown in the figure.

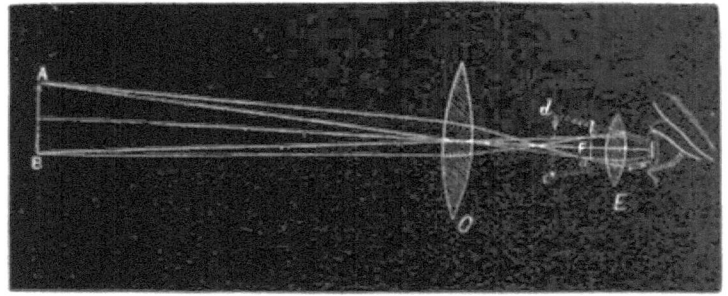

Fig. 300.

318. Astronomical Telescope. — The astronomical refracting telescope consists essentially, like the compound microscope, of two lenses. The object-glass (O, Fig. 300) forms a real diminished image *ab* of the object AB; this image, seen through the eye-glass E, appears magnified and of the size *cd*. The object-glass is of large diameter, in order to concentrate as much as possible the radiations from a distant object for a better illumination of the image.

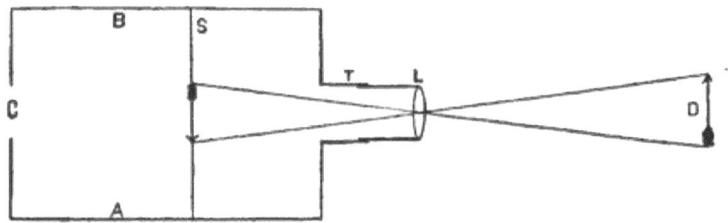

Fig. 301.

319. Photographer's Camera. — The *photographer's camera* or *camera obscura*, of which AB (Fig. 301) represents a vertical section, consists of a dark box painted black on the interior. A screen of ground glass S forms a partition in the box. A sliding tube T contains a convex lens L. If an object D is placed some distance in front, and the distance of the lens from the screen is suitably adjusted, a distinct, real, and inverted image can be seen upon the screen by looking through the aperture C. When the image is properly focused, the photographer replaces the ground-glass plate by a sensitized plate, and the chemical power of the sun's rays imprints a true picture of the object on this plate.

320. The Human Eye. — Figure 302 represents a horizontal section of this wonderful organ. Covering the front of the eye, like a watch-crystal, is a transparent coat 1, called the *cornea*. A tough membrane 2, of which the *cornea* is a continuation, forms the outer wall of the eye, and is called the *sclerotic coat*, or "white of the eye." This coat is lined on the interior with a delicate membrane 3, called the *choroid coat;* the latter contains a black pigment, which prevents internal reflection. The inmost coat 4, called the *retina*, is formed by expansion of the optic nerve O. The muscular tissue *ii* is called the *iris;* its color determines the so-called "color of the eye." In the center of the iris is a circular opening 5, called the *pupil*, whose function is to regulate, by involuntary enlargement and contraction, the quantity of light-waves admitted to the anterior chamber of the eye. Just back of the iris is a tough, elastic, and transparent body 6, called the *crystalline lens*. This lens divides the eye into two chambers; the anterior chamber 7 is filled with a limpid liquid, called the *aqueous humor;* the posterior chamber 8 is filled with a jelly-like substance, called the *vitreous humor*.

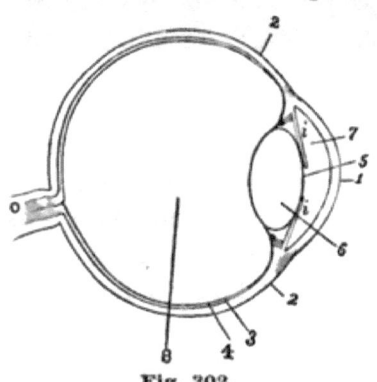

Fig. 302.

Experiment 280. — Make a model of an eye. Fill an 8-ounce flask with clear water (eye-ball). Cover one side with black paper having a round hole in it (iris and pupil). Place a slightly convex lens in front of the hole (cornea and crystalline lens combined; the latter outside the eye-ball instead of inside). Place a candle flame in front of the hole (object); catch (inverted) image of the flame on a paper screen (retina) behind the flask. Move the candle a little way from the flask; the image becomes indistinct. Restore it by interposing another convex lens (cure of long sight). Bring the candle near to the flask till the image is indistinct. Interpose concave lens to restore the clearness (cure of short sight).

Experiment 281. — Make two dots on paper two inches apart. Close the left eye, and bring the right one over the left spot. At a distance of about six inches the right spot becomes invisible. As you bring the paper nearer, the eye turns to regard the left spot, the image of the right spot meantime travels noseward over the retina, until it reaches a spot, called the *blind spot*, on the retina, which is not sensitive to the action of light-waves. This spot is where the optic nerve enters the eye.

The eye is a camera obscura, in which the retina serves as a screen. Images of outside objects are projected by means of the crystalline lens, assisted by the refractive powers of the humors, upon this screen, and the impressions thereby made on this delicate network of nerve filaments are conveyed by the optic nerve to the brain. If the two outer coatings are removed from the back part of the eye of an ox recently killed, so as to render it somewhat transparent, true images of whole landscapes may be seen formed upon the retina of the eye, when it is held in front of your eye. With the ordinary camera, the distance of the lens from the screen must be regulated to adapt itself to the varying distances of outside objects, in order that the images may be properly focused on the screen. In the eye this is accomplished by changing the convexity of the lens. We can almost instantly and involuntarily change the lens of the eye, so as to form on the retina a distinct image of an object miles away or only a few inches distant. The nearest limit at which an object can be placed, and form a distinct image on the retina, is about five inches. On the other hand, the normal eye in a passive state is adjusted for objects at an infinite distance.

Curiously enough, the retina, on careful examination, is found to be composed in part of little elements in its back portion, which have received, from their appearance, the names of *rods* and *cones*. It is thought that these rods and cones receive and respond to the vibrations of ether; in other words, that they co-vibrate with the undulations of the ether, and thereby we get our sensation of light.

321. Stereopticon. — This instrument is extensively employed in the lecture-room for producing on a screen magnified images of small, transparent pictures on glass,

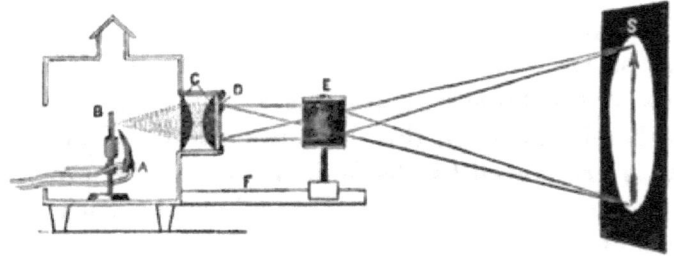

Fig. 303.

called *slides;* also for rendering a certain class of experiments visible to a large audience by projecting them on a screen. The *lime light* is most commonly used, though the electric light is preferred for a certain class of projections. The flame of an oxyhydrogen blow-pipe A (Fig. 303) is directed against a stick of lime B, and raises it to a white heat. The radiations from the lime are condensed, by means of a convex lens c, called the *condensing lens* (usually two plano-convex lenses are used), so that a larger quantity of radiations will pass through the convex lens E, called the *projecting lens*. The latter lens produces (or projects) a real, inverted, and magnified image of the picture on the screen S. The mounted lens

E may be slid back and forth on the bar F, so as properly to focus the image. (For useful information relating to the operation of projection, see Dolbear's Art of Projection.)

EXERCISES.

1. What is light?

2. State points of resemblance and points of difference between light-waves and sound-waves. Which can traverse a vacuum (as regards matter)?

3. Two books are held, respectively, 2 feet and 7 feet from the same gas-flame. Compare the intensities of the illumination of their respective pages.

4. What is the general effect of a concave mirror on light-waves? What kind of lens produces a similar effect?

5. How can a beam be bent?

6. State different ways by which the colors which compose white light may be revealed.

7. How do you account for the color of flowers? How do you account for the colors seen on a soap-bubble?

8. Why do white surfaces appear gray at twilight?

9. How are objects heated by the sun?

10. What evidences can you give that the earth receives energy from the sun?

APPENDIX.

Inches.

Millimeters. Centimeters.

The area of this figure is a square decimeter. A cube of water, one of whose sides is this area, is a cubic decimeter or a liter of water, and at the temperature of 4° C. weighs a kilogram. The same volume of air at 0° C., and under a pressure of one atmosphere, weighs 1.293 grams. The gram is the weight of 1^{cc} of pure water at 4° C.

Square Inch.

Square Centimeter

APPENDIX.

SECTION A.

Metric system of measures. — The term *metric* is derived from the word *meter*, which is the name of the fundamental unit employed in this system for measuring length, and from which all other units of the system are derived. The meter is, approximately, the ten-millionth part of the distance from the Equator to the North Pole. Defined by law, it is the distance at 0° C. between two lines engraved on a platinum bar kept in the Paris Observatory. The gram is theoretically the mass of 1^{cc} of distilled water at 4° C. By law it is $\frac{1}{1000}$ of the mass of a piece of platinum preserved in the same observatory. At Washington are kept exact copies of the meter and other metric measures.

The following tables contain all the requirements of this book. The pupil will find more complete tables in any good arithmetic.

TABLE OF LENGTHS.

10 millimeters (^{mm}) = 1 centimeter (^{cm}).
10 centimeters = 1 decimeter (^{dm}).
10 decimeters = 1 meter (^{m}).
1000 meters = 1 kilometer (^{km}).

TABLE OF AREAS.

100 square millimeters (^{qmm}) = 1 square centimeter (^{qcm}).
100 square centimeters = 1 square decimeter (^{qdm}).
100 square decimeters = 1 square meter (^{qm}).
1,000,000 square meters = 1 square kilometer (^{qkm}).

TABLE OF VOLUMES.

1000 cubic millimeters (^{cmm}) = 1 cubic centimeter (^{ccm} or ^{cc}).
1000 cubic centimeters = 1 cubic decimeter (^{cdm}).
1000 cubic decimeters = 1 cubic meter (^{cbm}).

The volumes of liquids and gases are either expressed in the units of the above table or in liters. The liter is 1^{cdm}, or 1000^{cc}.

TABLE OF MASSES OR WEIGHTS.

10 milligrams (^{mg}) = 1 centigram (^{cg}).
10 centigrams = 1 decigram (^{dg}).
10 decigrams = 1 gram (^{g}).
1000 grams = 1 kilogram or kilo (^{k}).

TABLE OF EQUIVALENTS.

1 inch = 0.0254 meter, or about $2\frac{1}{2}$ centimeters.
1 foot = 0.3048 meter, or about 30 centimeters.
1 yard = 0.9144 meter, or about $1\frac{9}{10}$ meter.
1 mile = 1609.0000 meters, or about $1\frac{6}{10}$ kilometers.

1 U.S. $\begin{cases} \text{liquid} \\ \text{dry} \end{cases}$ quart = $\begin{cases} 0.946 \text{ liter,} \\ 1.101 \text{ liters,} \end{cases}$ a little $\begin{cases} \text{less} \\ \text{more} \end{cases}$ than 1 liter.

1 U.S. gallon = 3.785 liters, or about $3\frac{8}{10}$ liters.

1 $\begin{cases} \text{avoirdupois} \\ \text{Troy and apothecaries'} \end{cases}$ ounce = $\begin{cases} 0.02835 \text{ kilo,} \\ 0.03110 \text{ kilo,} \end{cases}$ or rather $\begin{cases} \text{less} \\ \text{more} \end{cases}$ than 30 grams.

1 avoirdupois pound = 0.45359 kilo, or about $\frac{5}{11}$ kilo.

When great accuracy is not required, it will be found convenient to remember that

centimeters $\times \frac{2}{5}$ = inches (nearly);
inches $\times \frac{5}{2}$ = centimeters (nearly);
5 meters = 1 rod (nearly);
also, kilos $\times \frac{11}{5}$ = pounds (nearly);
pounds $\times \frac{5}{11}$ = kilos (nearly).

APPENDIX.

SECTION B.

TABLES OF SPECIFIC GRAVITIES OF BODIES.

[The standard employed in the tables of solids and liquids is distilled water at 4° C.]

I. *Solids.*

Antimony	6.712	Diamond	3.530
Bismuth	9.822	Glass, flint	3.400
Brass	8.380	Human body	0.890
Copper, cast	8.790	Ice	0.920
Iridium	23.000	Quartz	2.650
Iron, cast	7.210	Rock salt	2.257
Iron, bar	7.780	Saltpetre	1.900
Gold	19.360	Sulphur, native	2.033
Lead, cast	11.350	Tallow	0.942
Platinum	22.069	Wax	0.969
Silver, cast	10.470	Cork	0.240
Tin, cast	7.290	Pine	0.650
Zinc, cast	6.860	Oak	0.845
Anthracite coal	1.800	Beech	0.852
Bituminous coal	1.250	Ebony	1.187

II. *Liquids.*

Alcohol, absolute	0.800	Nitric acid	1.420
Bisulphide of carbon	1.293	Oil of turpentine	0.870
Ether	0.723	Olive oil	0.915
Hydrochloric acid	1.240	Sea water	1.026
Mercury	13.598	Sulphuric acid	1.841
Milk	1.032	Water, 4° C., distilled	1.000
Naphtha	0.847	Water, 0° C., distilled	0.999

III. *Gases.*

[Standard: air at 0° C.; barometer, 76^{cm}.]

Air	1.0000	Hydrogen	0.0693
Ammonia	0.5367	Nitrogen	0.9714
Carbonic acid	1.5290	Oxygen	1.1057
Chlorine	3.4400	Sulphuretted hydrogen	1.1912
Hydrochloric acid	1.2540	Sulphurous acid	2.2474

SECTION C.

TABLE OF NATURAL TANGENTS.

Deg.	Tangent.	Deg.	Tangent.	Deg.	Tangent.	Deg.	Tangent.
1	.017	24	.445	47	1.07	70	2.75
2	.035	25	.466	48	1.11	71	2.90
3	.052	26	.488	49	1.15	72	3.08
4	.070	27	.510	50	1.19	73	3.27
5	.087	28	.532	51	1.23	74	3.49
6	.105	29	.554	52	1.28	75	3.73
7	.123	30	.577	53	1.33	76	4.01
8	.141	31	.601	54	1.38	77	4.33
9	.158	32	.625	55	1.43	78	4.70
10	.176	33	.649	56	1.48	79	5.14
11	.194	34	.675	57	1.54	80	5.67
12	.213	35	.700	58	1.60	81	6.31
13	.231	36	.727	59	1.66	82	7.12
14	.249	37	.754	60	1.73	83	8.14
15	.268	38	.781	61	1.80	84	9.51
16	.287	39	.810	62	1.88	85	11.43
17	.306	40	.839	63	1.96	86	14.30
18	.325	41	.869	64	2.05	87	19.08
19	.344	42	.900	65	2.14	88	28.64
20	.364	43	.933	66	2.25	89	57.29
21	.384	44	.966	67	2.36	90	Infinite.
22	.404	45	1.000	68	2.48		
23	.424	46	1.036	69	2.61		

SECTION D.

REFERENCE TABLE OF RELATIVE RESISTANCES, ETC.

		Rel. Resist.	K.
Silver	@ 0° C.	1.00	9.15
Copper	"	1.06	9.72
Zinc	"	3.74	34.2
Platinum	"	6.02	55.1
Iron	"	6.46	59.1
German silver	"	13.91	127.3
Mercury	"	63.24	578.6

		Rel. Resist.
Nitric Acid — commercial	@ 15° to 28° C.	1,100,000
Sulphuric Acid, 1 to 12 parts water	"	2,000,000
Common salt — saturated sol.	"	3,200,000
Sulphate Copper "	"	18,000,000
Distilled water		not less than 10,000,000,000
Glass	@ 200° C.	15,000,000,000,000
Gutta percha	@ 0° C.	5,000,000,000,000,000,000

INDEX.

[NUMBERS REFER TO PAGES.]

A.

Aberration, Chromatic, 323.
   Spherical, 313.
Action and reaction, 14.
Adhesion, 25.
Air-pump, 41.
   Sprengel, 43.
Amalgamating battery zincs, 162.
Ampère, The, 172.
Ampère's rule, 158.
Ampère-volt, The, 172.
Atmosphere, 29.
Atmospheric pressure, measurement of, 33.

B.

Barometer, mercurial, 34.
   Aneroid, 35.
Batteries of different kinds, 164.
   of high resistance, 186.
   of low resistance, 186.
Batteries, Storage, 209.
Battery, what constitutes a voltaic, 185.
Beats in music, 261.
Boyle's or Mariotte's law, 40.
Buoyant force of fluids, 56.

C.

Calorie, The, 141.
Camera, Photographer's, 335.
Capillary phenomena, 27.
Celestial chemistry and physics, 322.
Center of gravity defined, 82.
   of gravity, how found, 83.
Centrifugal and centripetal forces, 92.
Cohesion, 20.
Cold, Methods of producing, artificially, 144.

Color, Cause of, revealed by dispersion. 317.
   produced by absorption, 324.
   produced by interference, 329.
Colors, Complementary, 329.
   Effect of contrast, 329.
   Effect of mixing, 326.
   Sky, 325.
Component forces, 72.
Composition of parallel forces, 75.
Compressibility of gases, 38.
Condenser, Air, 44.
Conduction of heat, 125.
Convection in gases, 126.
   in liquids, 129.
Coulomb, The, 171, 172.
Couple, Mechanical, 78.
Critical angle, 302.
Crystallization, 20.
Crystals, 21.
Curvilinear motion, 92.
Currents, Attraction and repulsion between, 192.
   Extra, 202.
   Induced, 200, 202.
   Laws of, 193, 194.
   Laws of induced, 202.
   Thermo-electric, 222.

D.

Density, 8, 59.
   Specific, 60.
Dew-point, 140.
Diathermancy and athermancy, 330.
Discord in music, 262.
Distillation, 138.
Divided circuits, 184.
Dynamo as an electric motor, 208.
   Uses of, 208.

Dynamo-electric machine, 205.
Dynamometers, 13.
Dyne, The, 106.
Ductility, 25.

E.

Ear, The, 279.
Elasticity, 24.
   of gases, 38.
Electrical measurements, 171.
Electric battery defined, 157.
Electrification, 226.
   confined to the external surface, 235.
   Two kinds of, 229.
Electric condenser, 234.
   current, chemical effects of, 166.
   current, heating and luminous effects of, 166.
   current, magnetic effects of, 170.
   current, physiological effects of, 169.
   current, direction of, 157.
   discharge, 231.
   energy, how it originates, 137.
   induction, 230.
   insulation, 232.
   machine, 232.
   motor, 204.
Electricity, Conductors and non-conductors of, 157.
   Static, 225.
   Two states of, 227.
Electro-chemical series, 161.
Electrolysis, 167.
Electro-motive force, 159.
   force of different batteries, 182.
Electrophorus, 233.
Electroplating and electrotyping, 214, 215.
Electroscope, 226.
Energy, 5.
   Distinction between force and, 102.
   Formulas for calculating kinetic, 103, 107.
   Kinetic and potential, 100.
   received from the sun, 281.
   Transformation, correlation, and conservation of, 147.
   Unit of, 101.
Equilibrant, 77.

Equilibrium, 13.
   of moments, 77.
   Three states of, 84.
Erg, The, 106.
Ether, a medium of motion, 282.
Ether-waves, Heating and chemical effects of, 322.
Evaporation, 139.
Expansion, Abnormal, 132.
   of solids, liquids, and gases, 130.
Eye, The human, 336.

F.

Falling bodies, laws of, 89.
   bodies, velocity of, independent of mass, 90.
Flexibility, 24.
Foot-pound, 101.
Fluids, 9.
Force, 11, 14.
   Centripetal and centrifugal, 92.
   Effect of a constant, 86.
   graphically represented, 70.
   how measured, 12.
   Moment of, 77.
Forces, Composition of, 72.
   Equilibrant of, 77.
   Resolution of, 73.
   Resultant of, 72.

G.

Galvanometer, 174.
   Tangent, 175.
   with astatic needle, 175.
Galvanoscope, 158.
Geissler tube, 203.
Gramme-dynamo, 205.
Gravitation and gravity, 15.
   Law of universal, 16.

H.

Hardening and annealing, 23.
Hardness, 22.
Heat, Artificial sources of, 122.
   generated by solidification and liquefaction, 143.
   Latent, 142.
   Mechanical equivalent of, 148.
   Theory of, 121.

INDEX. 351

Heat, The sun as a source of, 123.
    unit, 141.
Holtz machine, 160.
Horse-power, 105.
Hydrometers, 62.

I.

Images, 286.
    formed by lenses, 309.
    formed through apertures, 286.
    Virtual, 293.
Impenetrability, 2.
Incandescence, 283.
Induction coil, Ruhmkorff's, 202.
Induced currents, characteristics of, 204.
Inertia, 70.

J.

Joule's equivalent, 148.
    experiment, 147.

K.

Kinetic energy, 100.

L.

Lamp, Brush, 212.
    Electric, 211.
    Incandescent electric, 213.
Latent heat, 142.
Lenses, 305.
    Achromatic, 323.
    Effects of, 307.
Leyden jar, 234.
Light defined, 283.
    Electric, 210.
Lightning, 236.
    rods, 236.
Light-waves, Reflection of, 292.
    Sources of, 283.
    Velocity of, 292.
Liquefaction, 136.
Locomotive, The, 152.
Luminous and illuminated objects, 285.

M.

Machines, General law of, 111.
    Uses of, 108, 110.
Magnet, Ampère's theory of, 195.

Magnets, Coercive force of, 191.
    Forms of artificial, 192.
    Law of, 190.
    Polarity of, 191.
Magnetic equator, 198.
    field, 196.
    force, lines of, 196.
    needle, dip of, 198.
    needle, variation of, 198.
    poles of the earth, 197.
    transparency and induction, 190.
Malleability, 25.
Manipulation, 2.
Manometric flames, 268.
Mass, defined, 7.
Matter, Theory of its constitution, 7.
    What is it, 1.
Microphone, The, 221.
Microscope, Compound, 333.
    Simple, 312.
Minuteness of particles of matter, 6.
Mirrors, concave, 294.
    convex, 297.
    plane, 293.
Mixing colors, effects of, 326.
    pigments, effects of, 328.
Molar forces, 19.
Molecular forces, 18, 19.
Moment of a force, 77.
Momentum, 67.
    its relation to force, 67.
Motion, First Law of, 69.
    Graphical representation of, 70.
    Relative, 10.
    Second Law of, 71.
    Third Law of, 80.
Musical instruments, 270.
    scale, 259.

N.

Nodes, 240.

O.

Ohm, 174, 178.
Ohm's law, 182.
Overtones and harmonics, 262.

P.

Pendulum, Laws of, 95.
Phonograph, The, 277.

Phosphorescence, 283.
Photometry, 289.
Physics defined, 1.
Pitch, Musical, 259.
Polarization of electric elements, 164.
Pores and porosity, 7.
Potential, Electric, 159.
Press, Hydrostatic, 50.
Pressure, Atmospheric, 29.
  in fluids, 29, 51.
  transmitted by fluids, 47.
Prisms, Optical, 305.
Pump, Air, 41.
  Force, 46.
  Lifting or suction, 44.
Pump, Sprengel, 43.

Q.

Quality of sound, 265.

R.

Radiant energy, 281.
Radiation, 129.
  Only one kind of, 322.
  Thermal effects of, 330.
Radiometer, 281.
Rainbow, The, 315.
Ray, beam, and pencil defined, 284.
Reflection, Total, 302.
Refraction, 298.
  Cause of, 300.
  Double, 304.
  Index of, 300, 301.
Relay and repeater, 217.
Resistance measured by substitution, 179.
  of battery, 178.
  of electric conductors, 176.
Resonators, 253.
Rheostat, Description of, 178.

S.

Shadows, 287.
Shunts, 184.
Siphon, 54.
Sonometer, 260.
Sound, Analysis of, 265.
  defined, 247.

Sound, Intensity of, 251.
  Quality of, 265.
  Synthesis of, 266.
Sounding-plates and bells, 273–275.
Sound-vibrations, Method of representing graphically, 267.
Sound-waves, How they originate, 244.
  How they travel, 245.
  Measuring length and velocity of, 255
  Media for transmitting, 247.
  Reënforcement and interference of, 253, 256
  Reflection of, 249.
  Velocity of, 248.
Speaking-tubes, 252.
Specific gravity and specific density, 59.
  Formulas for, 60.
Spectra, 314.
  Bright-line, 318.
  Dark-line, 320.
  Continuous, 318.
Spectrum analysis, 321.
Stability of a body, on what it depends, 85.
Steam-engine, Compound, 152.
  Condensing and non-condensing, 151.
  Description of simple, 149.
Stereopticon, 338.
Storage batteries, 209.
Surface of a liquid at rest is level, 53.
Synthesis of white waves, 316.

T.

Telescope, Astronomical, 335.
Telegraph, The, 216.
Telephone, The Bell, 218.
Temperature, defined, 124.
  distinguished from quantity of heat, 124.
Temperatures, Standard, 133.
Tenacity, 20.
Tension, 26.
Theory of exchanges, 332.
Thermo-dynamics defined, 147.
Thermo-electric batteries and thermopiles, 224.
  currents, 222.
  series, 224.
Thermometer, Construction of, 133
  Graduation of, 133.

Thermometry, 133.
Three states of matter, 9.
Transformation of electric energy, 208.
    of electric energy into heat, 189.
    of heat energy into electric energy, 222.
    of mechanical energy into electric potential energy, 225.
Transparency, translucency, and opacity, 285.

U.

Undulatory theory of radiation, 283.
Unit of heat, 141.
    of intensity of a magnetic field, 173.
    of magnetic pole, 173.
Units, Absolute, 106.
    C.G.S. magnetic and electro-magnetic, 173.
    Fundamental and derived, 106.

V.

Vaporization, 136.
Ventilation, 128.
Vibration of strings, 260.
    Period of, 238.
Vibrations forced and sympathetic, 257.
    Graphical method of studying, 243.
    Propagation of, 239.
    Stationary, 240.

Viscosity, 24.
Visual angle, 291.
Vocal organs, 276.
Volt, The, 172.
Voltaic arc, 210.
    cells, best arrangement of, 187.
    cells connected in opposition, 185.
    cells, methods of combining, 185.
Volume, 7.

W.

Watt, 172.
Waves, 239.
    Amplitude of, 239.
    how propagated, 243
    Interference of, 239.
    Length of, 239.
    Longitudinal and transverse, 241, 242.
    Reflection of, 239.
Wave-motion, Air as a medium of, 242.
Weight, 16.
    Point of maximum, 17.
Welding, 20.
Wheatstone bridge, 180.
Work, 98.
    Formula for estimating, 99.
    Rate of doing, 104.
    Unit of, 101.
    wasted, 103.

SCIENCE AND HISTORY.

NATURAL SCIENCE.

		INTROD. PRICE
Everett:	Vibratory Motion and Sound	$2.00
Gage:	Elements of Physics	1.12
	Introduction to Physical Science	1.00
Hale:	Little Flower-People	.40
Hill:	Questions on Stewart's Physics	.35
Journal of Morphology	(per vol.)	6.00
Knight:	Primer of Botany	.30
Williams:	Introduction to Chemical Science	.80

PHILOSOPHICAL SCIENCE.

Davidson:	Rosmini's Philosophical System	2.50
Hickok:	Philosophical Works	.90
Ladd:	Lotze's Outlines of Metaphysic	.80
	Lotze's Outlines of Philosophy of Religion	.80
	Lotze's Outlines of Practical Philosophy	.80
	Lotze's Outlines of Psychology	.80
	Lotze's Outlines of Æsthetics	.80
	Lotze's Outlines of Logic	.80
Seelye:	Hickok's Mental Science (Empirical Psychology)	1.12
	Hickok's Moral Science	1.12

POLITICAL SCIENCE.

Clark:	Philosophy of Wealth		1.00
Clark & Giddings:	The Modern Distributive Process	(retail)	.75
Macy:	Our Government		.70
Political Science Quarterly		(per vol.)	3.00
Seligman:	Railway Tariffs and the Interstate Law	(retail)	.75

HISTORY.

Allen:	Readers' Guide to English History		.25
Andrade:	Historia do Brazil		.75
Fiske-Irving:	Washington and His Country		1.00
Halsey:	Genealogical and Chronological Chart		.25
Journal of Archæology		(per vol.)	5.00
Judson:	Cæsar's Army		1.60
Montgomery:	Leading Facts of English History		1.00
	English History Reader		.60
Moore:	Pilgrims and Puritans		.60
Myers:	Mediæval and Modern History		1.50
	Ancient History		1.40

Copies sent to Teachers for Examination, with a view to Introduction, on receipt of Introduction Price.

GINN & COMPANY, Publishers.

BOSTON.   NEW YORK.   CHICAGO.

BOOKS ON ENGLISH LITERATURE.

Allen	Reader's Guide to English History	$ .25
Arnold . . .	English Literature	1.50
Bancroft . .	A Method of English Composition	.50
Browne . .	Shakespere Versification	.25
Fulton & Trueblood:	Choice Readings	1.50
	Chart Illustrating Principles of Vocal Expression,	2.00
Genung . .	Practical Elements of Rhetoric	1.25
Gilmore . .	Outlines of the Art of Expression	.60
Ginn	Scott's Lady of the Lake . . . *Bds.*, .35; *Cloth*,	.50
	Scott's Tales of a Grandfather . *Bds.*, .40; *Cloth*,	.50
Gummere .	Handbook of Poetics	1.00
Hudson . .	Harvard Edition of Shakespeare : —	
	20 Vol. Edition. *Cloth, retail*	25.00
	10 Vol. Edition. *Cloth, retail*	20.00
	Life, Art, and Character of Shakespeare. 2 vols.	
	Cloth, retail	4.00
	New School Shakespeare. *Cloth.* Each Play .	.45
	Old School Shakespeare, per play	.20
	Expurgated Family Shakespeare	10.00
	Essays on Education, English Studies, etc. . .	.25
	Three Volume Shakespeare, per vol.	1.25
	Text-Book of Poetry	1.25
	Text-Book of Prose	1.25
	Pamphlet Selections, Prose and Poetry . . .	.15
	Classical English Reader	1.00
Johnson . .	Rasselas . . . . *Bds.*, .30; *Cloth*,	.40
Lee	Graphic Chart of English Literature	.25
Martineau .	The Peasant and the Prince . *Bds.*, .35; *Cloth*,	.50
Minto . . .	Manual of English Prose Literature	1.50
	Characteristics of English Poets	2.00
Rolfe	Craik's English of Shakespeare	.90
Scott	Guy Mannering . . . . . *Bds.*, .60; *Cloth*,	.75
	Ivanhoe . . . . *Bds.*, .60; *Cloth*,	.75
	Talisman . . . . *Bds.*, .50; *Cloth*,	.60
	Rob Roy . . . . *Bds.*, .60; *Cloth*,	.75
Sprague .	Milton's Paradise Lost, and Lycidas	.45
	Six Selections from Irving's Sketch-Book	
	Bds., .25; *Cloth*,	.35
Swift . . .	Gulliver's Travels . . . . *Bds.*, .30; *Cloth*,	.40
Thom . . .	Shakespeare and Chaucer Examinations . . .	.00

Copies sent to Teachers for Examination, with a view to Introduction,
on receipt of the Introduction Price given above.

GINN & COMPANY, Publishers,
Boston, New York, and Chicago.

Greek Text-Books.

		Intro. Price
Allen:	Medea of Euripides	$1.00
Flagg:	Hellenic Orations of Demosthenes	1.00
	Seven against Thebes	1.00
	Anacreontics	.35
Goodwin:	Greek Grammar	1.50
	Greek Reader	1.50
	Greek Moods and Tenses	1.50
	Selections from Xenophon and Herodotus	1.50
Goodwin & White	Anabasis, with vocabulary	1.50
Harding:	Greek Inflection	.50
Keep:	Essential Uses of the Moods	.25
Leighton:	New Greek Lessons	1.20
Liddell & Scott:	Abridged Greek-English Lexicon	1.90
	Unabridged Greek-English Lexicon	9.40
Parsons:	Cebes' Tablet	.75
Seymour:	Selected Odes of Pindar	1.40
	Introd. to Language and Verse of Homer, { Paper	.45
	{ Cloth	.60
Sidgwick:	Greek Prose Composition	1.50
Tarbell:	Philippics of Demosthenes	1.00
Tyler:	Selections from Greek Lyric Poets	1.00
White:	First Lessons in Greek	1.20
	Schmidt's Rhythmic and Metric	2.50
	Oedipus Tyrannus of Sophocles	1.12
	Stein's Dialect of Herodotus	.10
Whiton:	Orations of Lysias	1.00

College Series.

- **Beckwith**: Euripides' Bacchantes.
  Text and Notes, Paper, .80; Cloth, $1.10; Text only, .20.
- **D'Ooge**: Sophocles' Antigone.
  Text and Notes, Paper, .95; Cloth, $1.25; Text only, .20.
- **Dyer**: Plato's Apology and Crito.
  Text and Notes, Paper, .95; Cloth, $1.25; Text only, .20.
- **Fowler**: Thucydides, Book V.
  Text and Notes, Paper, .95; Cloth, $1.25; Text only, .20.
- **Humphreys**: Aristophanes' Clouds.
  Text and Notes, Paper, .95; Cloth, $1.25; Text only, .20.
- **Manatt**: Xenophon's Hellenica, Books I.-IV.
  Text and Notes, Paper, $1.20; Cloth, $1.50; Text only, .20.
- **Morris**: Thucydides, Book I.
  Text and Notes, Paper, $1.20; Cloth, $1.50; Text only, .20.
- **Seymour**: Homer's Iliad, Books I.-III.
  Text and Notes, Paper, .95; Cloth, $1.25; Text only, .20.
- **Smith**: Thucydides, Book VII.
  Text and Notes, Paper, .95; Cloth, $1.25; Text only, .20.

Sanskrit.

Arrowsmith:	Kaegi's Rigveda, (*translation*)	$1.50
Elwell:	Nine Jatakas (*Pali*)	.50
Lanman:	Sanskrit Reader	1.80
Perry:	Sanskrit Primer	1.50
Whitney:	Sanskrit Grammar	2.50

Copies sent to Teachers for Examination, with a view to Introduction, on receipt of Introduction Price.

GINN & COMPANY, Publishers,

BOSTON, NEW YORK, AND CHICAGO.

MODERN LANGUAGES, Etc

MODERN LANGUAGES.

		IROD. PRICE
Becker:	Spanish Idioms	$1.80
Collar:	Eysenbach's German Lessons	1.20
Cook:	Table of German Prefixes and Suffixes	.05
Doriot:	Illustrated Beginners' Book in French	.80
Knapp:	Modern French Readings	.80
	Modern Spanish Readings	1.50
	Modern Spanish Grammar	1.50
Spiers:	French-English General Dictionary	4.50
	English-French General Dictionary	4.50
Stein:	German Exercises	.40

MISCELLANEOUS.

Ariel:	Those Dreadful Mouse Boys	.75
Arrowsmith:	Kaegi's Rigveda	1.50
Burgess:	The American University	.15
Channing:	Delbrück's Introduction to the Study of Language	.94
Culver:	Epitome of Anatomy	.20
Davidson:	The Place of Art in Education	.20
	The Parthenon Frieze and other Essays	1.50
	A Dante Handbook	1.12
Dippold:	Emanuel Geibel's Brunhild	.45
Fisk:	Teachers' Improved Class Books	
Flagg:	Pedantic Versicles	.75
Halsey:	Genealogical and Chronological Chart	.25
Harrington:	Helps to the Intelligent Study of College Preparatory Latin	
Hitchcock:	Manual for the Gymnasium	.25
Hofmann:	Inaugural Address	.25
Hudson:	Daniel Webster	.25
J. B. G.:	Queen of Hearts	.20
Lectures on School Hygiene		.80
Leighton:	Harvard Examination Papers	1.00
Prince:	Courses of Studies and Methods of Teaching	.75
Monoyer:	Sight Test	.12
Packard:	Studies in Greek Thought	.80
Seelye:	Growth through Obedience	.25
Smith & Blackwell:	Parallel Syntax Chart	1.00
Stevens:	Yale Examination Papers	.75
Straight:	True Aim of Industrial Education	.10
Super:	Weil's Order of Words in the Ancient Languages compared with the Modern	1.12
Warren:	True Key to Ancient Cosmology	.20

Copies sent to Teachers for Examination, with a view to Introduction, on receipt of Introduction Price.

GINN & COMPANY, Publishers.

BOSTON.   NEW YORK.   CHICAGO.

Latin Text-Books.

		INTROD. PRICE.
ALLEN & GREENOUGH:	Latin Grammar	$1.20
	Cæsar (7 books, with vocabulary; illustrated)	1.25
	Cicero (13 orations, with vocabulary; illustrated)	1.25
	Sallust's Catiline	.60
	Cicero de Senectute	.50
	Ovid (with vocabulary)	1.40
	Preparatory Course of Latin Prose	1.40
	Latin Composition	1.12
ALLEN	New Latin Method	.90
	Introduction to Latin Composition	.90
	Latin Primer	.90
	Latin Lexicon	.90
	Remnants of Early Latin	.75
	Germania and Agricola of Tacitus	1.00
BLACKBURN	Essentials of Latin Grammar	.70
	Latin Exercises	.60
	Latin Grammar and Exercises (in one volume)	1.00
COLLAR & DANIELL:	Beginner's Latin Book	1.00
	Latine Reddenda (paper)	.20
	Latine Reddenda and Voc. (cloth)	.30
COLLEGE SERIES OF LATIN AUTHORS.		
	Greenough's Satires and Epistles of Horace	
	(text edition) $0.20; (text and notes)	1.25
CROWELL	Selections from the Latin Poets	1.40
CROWELL & RICHARDSON:	Brief History of Roman Lit. (BENDER)	1.00
GREENOUGH	Virgil:—	
	Bucolics and 6 Books of Æneid (with vocab.)	1.60
	Bucolics and 6 Books of Æneid (without vocab.)	1.12
	Last 6 Books of Æneid, and Georgics (with notes)	1.12
	Bucolics, Æneid, and Georgics (complete, with notes)	1.60
	Text of Virgil (complete)	.75
	Vocabulary to the whole of Virgil	1.00
GINN & CO.	Classical Atlas and Geography (cloth)	2.00
HALSEY	Etymology of Latin and Greek	1.12
KEEP	Essential Uses of the Moods in Greek and Latin	.25
KING	Latin Pronunciation	.25
LEIGHTON	Latin Lessons	1.12
	First Steps in Latin	1.12
MADVIG	Latin Grammar (by THACHER)	2.25
PARKER & PREBLE:	Handbook of Latin Writing	.50
PREBLE	Terence's Adelphoe	.25
SHUMWAY	Latin Synonymes	.30
STICKNEY	Cicero de Natura Deorum	1.40
TETLOW	Inductive Latin Lessons	1.12
TOMLINSON	Manual for the Study of Latin Grammar	.20
	Latin for Sight Reading	1.00
WHITE (J. W.)	Schmidt's Rhythmic and Metric	2.50
WHITE (J. T.)	Junior Students' Latin-English Lexicon (mor.)	1.75
	English-Latin Lexicon (sheep)	1.50
	Latin-English and English-Latin Lexicon (sheep)	3.00
WHITON	Auxilia Vergiliana; or, First Steps in Latin Prosody	.15
	Six Weeks' Preparation for Reading Cæsar	.40

Copies sent to Teachers for Examination, with a view to Introduction, on receipt of Introduction Price.

GINN & COMPANY, Publishers,
BOSTON, NEW YORK, AND CHICAGO.

Mathematics.[2]

		Introd. Prices.
Byerly	Differential Calculus	$2.00
	Integral Calculus	2.00
Ginn	Addition Manual	.15
Halsted	Mensuration	1.00
Hardy	Quaternions	2.00
Hill	Geometry for Beginners	1.00
Sprague	Rapid Addition	.10
Taylor	Elements of the Calculus	1.80
Wentworth	Grammar School Arithmetic	.75
	Shorter Course in Algebra	1.00
	Elements of Algebra	1.12
	Complete Algebra	1.40
	Plane Geometry	.75
	Plane and Solid Geometry	1.25
	Plane and Solid Geometry, and Trigonometry	1.40
	Plane Trigonometry and Tables. *Paper*	.60
	Pl. and Sph. Trig., Surv., and Navigation	1.12
	Pl. and Sph. Trig., Surv., and Tables	1.25
	Trigonometric Formulas	1.00
Wentworth & Hill:	Practical Arithmetic	1.00
	Abridged Practical Arithmetic	.75
	Exercises in Arithmetic	
	Part I. *Exercise Manual*	
	Part II. *Examination Manual*	.35
	Answers (to both Parts)	.25
	Exercises in Algebra	.70
	Part I. *Exercise Manual*	.35
	Part II. *Examination Manual*	.35
	Answers (to both Parts)	.25
	Exercises in Geometry	.70
	Five-place Log. and Trig. Tables (7 *Tables*)	.50
	Five-place Log. and Trig. Tables (*Comp. Ed.*)	1.00
Wentworth & Reed:	First Steps in Number, *Pupils' Edition*	.30
	Teachers' Edition, complete	.90
	Parts I., II., and III. (separate), each	.30
Wheeler	Plane and Spherical Trig. and Tables	1.00

Copies sent to Teachers for examination, with a view to Introduction, on receipt of Introduction Price.

GINN & COMPANY, Publishers.

BOSTON.   NEW YORK.   CHICAGO.

www.ingramcontent.com/pod-product-compliance
Lightning Source LLC
Chambersburg PA
CBHW031419230426
43668CB00007B/368